Franz Rothe

A Course in Old and New Geometry

Volume II:
Basic Euclidean Geometry

Preface

The *front cover* shows the figure from page 113 about the geometric proof of the law $a(bc) = b(ac)$. This line of reasoning does not presuppose Pappus' theorem, instead we use properties of circles in a Pythagorean plane.

A decade long experience of teaching the course *Fundamental of Geometry* at the University of North Carolina at Charlotte, many notes for exercises, reading and research are the bases for this bulky work. The present second volume starts with Euclidean geometry, and the text gets a more educational and even more elementary flavour. Here I start with Thales theorem about the angle in a semicircle, and continue with Euclid's related theorems about angles in a circle. The most simple parts of the Euclidean geometry are given in detail, as well as the later parts about similarity, and finally area in Euclidean geometry with the theorem of Pythagoras and trigonometry, and the measurement of the circle. Too, the lens equation from geometrical optics is treated.

In several sections is recalled the strictly modern view of the first volume, beginning with Hilbert's axioms from the *Foundations of Geometry* [9]. After some discussion of logic and axioms in general, we go on with a short section about incidence geometries in two dimensions. The theorems from neutral geometry which have been proved in the first volume and are needed again, are cited. The relation of analytic and synthetic geometry is treated on Hilbert's rigorous account.

Retired I could find the time and leisure to prepare several volumes. My heartfelt thanks go to everyone involved, this means famous authors as well as my former students.

Charlotte, November 2023.

MSC2020-Mathematics Subject Classification

51-01 Introductory exposition (textbooks, tutorial papers, etc.) pertaining to geometry

51-02 Research exposition (monographs, survey articles) pertaining to geometry

51-03 History of geometry

51F20 Congruence and orthogonality in metric geometry

51N20 Euclidean analytic geometry

About the author:

Franz Rothe graduated from high school in Karlsruhe and studied mathematics, physics and music there. Graduated with a diploma in mathematics from the E T H Zürich, a doctorate in Tübingen. He got his Habilitation and venia legendi from the University at Tübingen (1984) and the Ludwig Maximilian University of Munich

(1988). He has published about 40 articles, and meanwhile several books about mathematics. 1'p[;21Since 1990 a professorship at the University of North Carolina at Charlotte, USA. His is now retired because of health problems.

In addition, Rothe and pianist Thomas Turner have developed a repertoire of classical music for flute and piano, and have recorded and released three CDs. This collection also contains several of their own transcriptions.

Some recent information about activity, in all fields, is gathered on my website.

Contents

Part I

First Steps in Euclidean Geometry

I.1 Some Euclidean Geometry of Circles

Some interesting properties of circles and rectangles in Euclidean geometry are investigated in this section. The material is from book III of Euclid, and includes furthermore the converse of Thales' theorem, and the basic properties of rectangles. Recall that in Euclidean geometry, the parallel axiom and its consequences are now assumed to hold.

I.1.1 Thales' Theorem

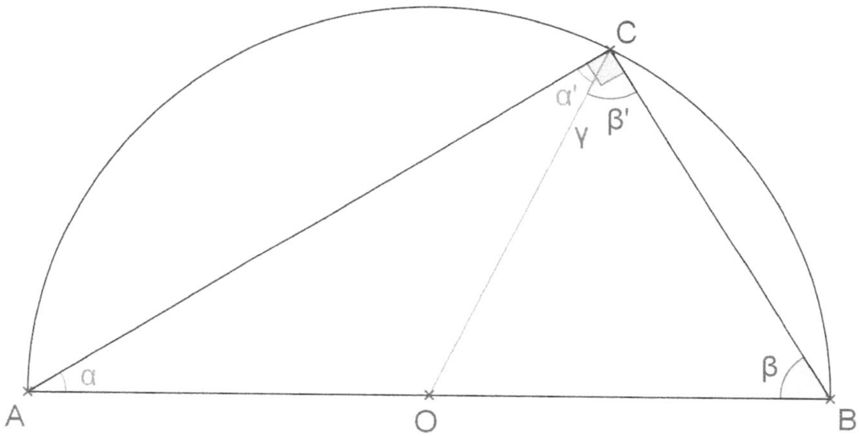

Figure I.1.1: Thales Theorem.

Theorem I.1.1 (Thales' Theorem). *"The angle in a semicircle is a right angle." More precisely stated: If an angle has its vertex C on a circle, and its sides cut the circle at the two endpoints A and B of a diameter, then angle $\gamma = \angle ACB$ is a right angle.*

Thales's lived ca. 624-546 B.C., in Miletus, a Greek island along the coast of Asia Minor. These dates are known rather precisely, because, as reported by Herodotus, he predicted a solar eclipse, which has been determined by modern methods to have occurred on May 28th of 585 B.C.

Tradition names Thales of Miletus as the first Greek philosopher, mathematician and scientist. He is known for his theoretical as well as practical understanding of geometry. Most important, he is credited with introducing the concept of logical proof for abstract propositions.

Traditions surrounds him with legends. Herodotos mentioned him as having predicted a solar eclipse, which put an end to fighting between the Lydians and the Medes.

Aristotle tells this story about him:

> Once Thales somehow deduced that there would be a great harvest of olives in the coming year; so, having a little money, he gave deposits for the use of all the olive presses in Chios and Miletus, which he hired at a low price because no one bid against him. When the harvest-time came, and many were wanted all at once and of a sudden, he let them out at any rate he pleased, and made a quantity of money.

Plutarch tells the following story:

> Solon who visited Thales asked him the reason which kept him single. Thales answered that he did not like the idea of having to worry about children. Nevertheless, several years later Thales anxious for family adopted his nephew Cybisthus.

Thales went to Egypt and studied with the priests. While he was in Egypt, he was able to determine the height of a pyramid by measuring the length of its shadow at the moment when the length of his own shadow was equal to his height. Thales is said to have proved some simple theorems of geometry, as well as the not so obvious theorem about the right angle in a semicircle. As stressed by David Park [15], the story raises an important point, whether or not Thales really invented the proof:

> Babylonians and Egyptians had a number of mathematical tricks. For example, Babylonians knew this proposition, as well as the Pythagorean theorem a thousand years before Thales and Pythagoras found them. If they were known, they must have been proved, but there is no sign that anyone thought the proofs were important enough to preserve. Whoever set the process of proof at the center of the stage is the founder of all the mathematics since then, and if it is not Thales it was someone who lived not long afterwards.

Proof of Thales' Theorem in modern manners. Draw $\triangle ABC$ and, as an extra for the proof, the line from the center O to the vertex C. The base angles of an isosceles triangle are equal by Euclid I.5 . In modern parlance, we say:

`The base angles of an isosceles triangle are congruent.`

We use that fact at first for $\triangle AOC$. Hence

$$\alpha = \angle OAC \cong \angle OCA$$

Secondly, we use Euclid I.5 for $\triangle COB$. Hence

$$\beta = \angle OBC \cong \angle OCB$$

By angle addition at vertex C

$$\gamma = \angle ACB = \alpha + \beta \tag{I.1.1}$$

Next we use Euclid I.32, which tells us:

`The sum of the angles in a triangle is two right angles.`

Because α, β, γ are just the angles in $\triangle ABC$, we conclude that

$$\alpha + \beta + \gamma = 2R \tag{I.1.2}$$

I have used the letter R to denote a right angle. Pulling formulas (I.1.1) and (V.12.1) together yields $\gamma + \gamma = \alpha + \beta + \gamma = 2R$, and hence $\gamma = R$, as to be shown. □

Problem I.1.1 (A special triangle). *Use the equilateral triangle and Thales' Theorem to construct a triangle with angles of* $30°, 60°$ *and* $90°$.

Answer. On an arbitrary segment AB, an equilateral triangle is erected. This can be done as in Euclid I.1 by finding the intersection point C of a circle around A through B with a circle around B through A. We extend the segment AB on the side opposite to B, and get the second intersection point D of the extension with the circle around A. The triangle $\triangle DBC$ has the angles $30°, 60°$ and $90°$ at its vertices D, B and C.

Problem I.1.2. *Why are all three triangles in the figure on page 6 equilateral. To give a valid reason, use only the following facts:*

- *The base angles of a triangle are congruent if and only if the triangle is isosceles.*
- *The sum of the angles in a triangle is* $180°$.

Proof. By construction, the right-hand triangle $\triangle AMD$ and the left-hand triangle $\triangle MBC$ are both equilateral. Both have three congruent angles. Because of the angle sum, their angles are all $60°$. Angle addition at the vertex M implies $\angle CMD = 180° - 60° - 60° = 60°$. Moreover, the middle triangle $\triangle MCD$ is isosceles by construction. Hence the two base angles marked β are congruent. Finally the angle sum of the middle triangle implies $60° + 2\beta = 180°$, and hence $\beta = 60°$. We see that the middle triangle has three angles of $60°$, too. Hence all its three sides are congruent, and it is equilateral, too. □

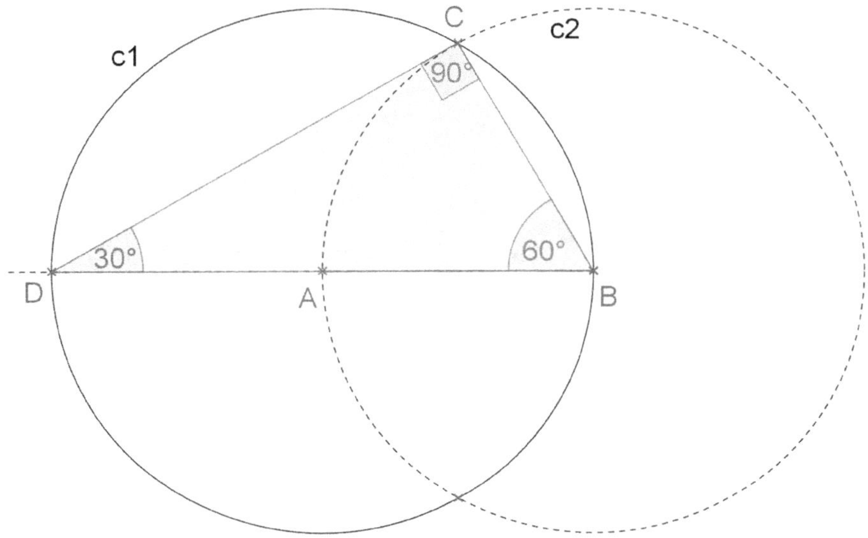

Figure I.1.2: Construction of a triangle with angles of $30°, 60°$ and $90°$.

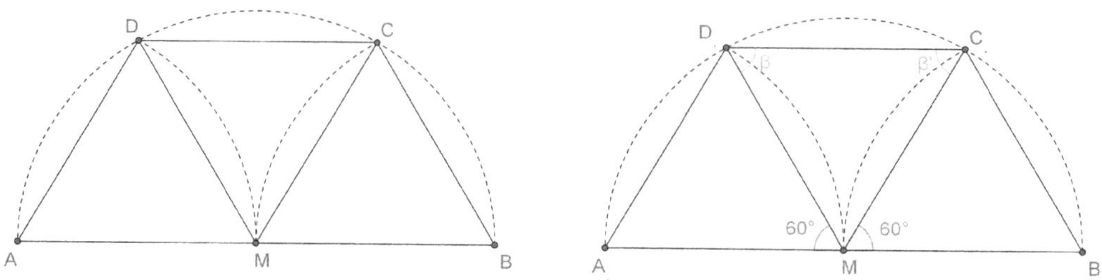

Figure I.1.3: Three triangles.

Problem I.1.3. *Construct a right triangle with hypothenuse $c = 5$ and one leg $a = 3$. Use a construction based on Thales' theorem and describe your construction.*

Answer. We draw a segment of the length $|AB| = c = 5$ as given. Draw a semicircle with diameter AB. Draw a circle around B with radius 3. The semicircle and the circle intersect at point C. The triangle $\triangle ABC$ is a right triangle. Hypothenuse $|AB| = 5$, and leg $|BC| = 3$ have the lengths as required.

Problem I.1.4. *Construct a right triangle with projections $p = 3$ and $q = 4$ of the legs onto the hypothenuse. Use a construction based on Thales' theorem and describe*

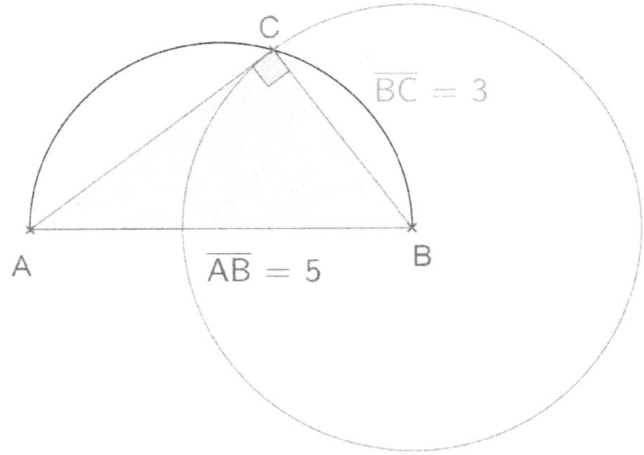

Figure I.1.4: Construction of a right triangle with hypothenuse $c = 5$ and leg $a = 3$.

your construction.

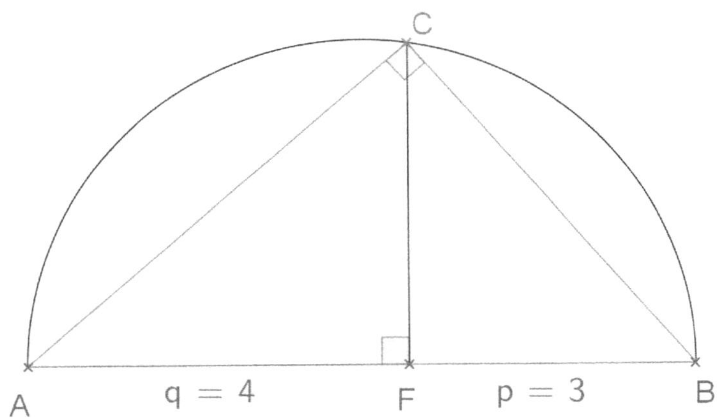

Figure I.1.5: Construction of a right triangle with projections $p = 3, q = 4$.

Answer. We draw segments of the lengths as given, $|AF| = q = 4$ and $|FB| = p = 3$, adjacent to each other on one line. Erect the perpendicular on line AB at point F. Draw a semicircle with diameter AB. The semicircle and the perpendicular intersect at point C. The triangle $\triangle ABC$ is a right triangle with hypothenuse AB, and the projections $q = AF$ and $FB = p$ have the lengths as required.

Problem I.1.5. *Use Thales' theorem for a Euclidean construction of the perpendicular p to a given line l through a given point P lying on l. Do and describe the construction.*

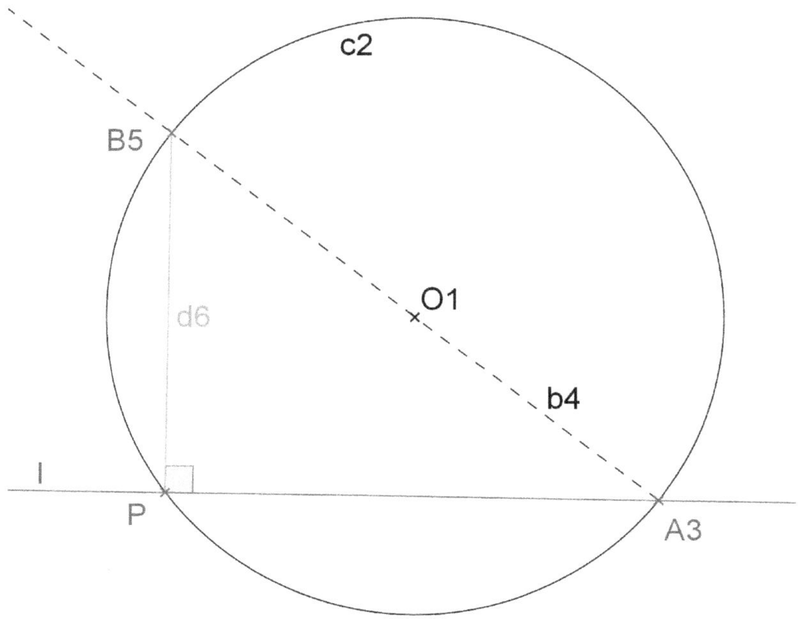

Figure I.1.6: Erect the perpendicular via Thales' Theorem. The numbers indicate the order of the steps.

Answer. One chooses an arbitrary point O not on line l, and draws a circle c around it through point P. This circle intersects the given line l at point P, and a second point, which is called A. Next one draws the line OA. It intersects the circle c at A and a second point, called B. Finally, the line BP is the perpendicular to be erected on line l at point P.

Problem I.1.6. *Use Thales' Theorem to drop the perpendicular from a given point P not lying on line l onto the given line l.*

Answer. The construction is indeed a bid awkward—and done more for fun than of practical value.

Theorem I.1.2 (A strengthening of Thales' Theorem). *Given is a triangle $\triangle ABC$, and a circle with its side AB as diameter.*

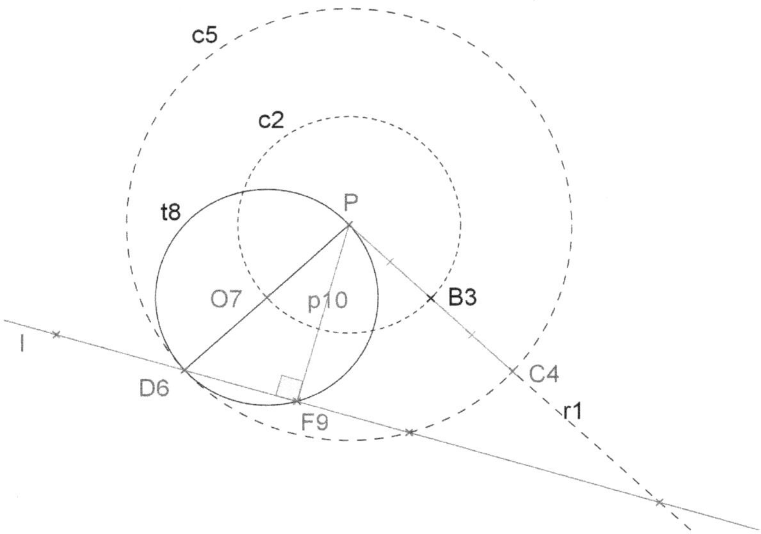

Figure I.1.7: Drop the perpendicular via Thales' Theorem. The numbers indicate the order of the steps.

(i) *If the third vertex C of the triangle lies inside the circle, the angle at vertex C is obtuse.*

(ii) *If the third vertex C of the triangle lies outside the circle, the angle at vertex C is acute.*

(iii) *If the third vertex C of the triangle lies on the circle, the angle at vertex C is right.*

Corollary 1 (Converse Thales' Theorem). *If the diameter of a circle is one side of a right triangle, then its third vertex lies on this circle, too.*

Theorem I.1.3 (Converse Thales' Theorem). *The vertex of a right angle the sides of which cut a circle at the endpoints of a diameter, lies on this circle.*

Taking Thales' Theorem I.1.1 and the converse Theorem I.1.3 together, we can state:

Corollary 2 (Strong Thales' Theorem). *An angle the sides of which cut a circle at the endpoints of a diameter is a right angle if and only if its vertex lies on this circle.*

Proof of Theorem I.1.2. Following Euler's convention, let $\alpha = \angle BAC$, $\beta = \angle ABC$ and $\gamma = \angle ACB$ denote the angles at the vertices A, B and C, respectively. Let O be the midpoint of the triangle side AB. In the figures on page 11 and page 11, there are given drawings of the $\triangle ABC$ and the circle for the two cases:

(i) $OC < OA$ or

(ii) $OC > OA$.

We use Euclid I.18:

> `If one side of a triangle is greater than another, then the angle opposite to the greater side is greater than the angle opposite to the smaller side.`

First consider the case that (i) $OC < OA$ holds. Using Euclid I.18 for $\triangle AOC$, we conclude that

$$\alpha < \alpha' = \angle OCA \tag{I.1.3}$$

Now $OA \cong OB$, since O is the midpoint of the hypothenuse AB. Hence, because of (i), $OC < OB$ holds, too. Using Euclid I.18 once more, this time for triangle $\triangle BOC$, we conclude that

$$\beta < \beta' = \angle OCB \tag{I.1.4}$$

Adding (I.1.3) and (I.1.4) yields $\alpha + \beta < \alpha' + \beta'$. Angle addition at vertex C yields $\alpha' + \beta' = \gamma$. Now we use that the sum of the angles in a triangle is two right angles, as stated in Euclid I.32. Hence

$$2R = \alpha + \beta + \gamma < \alpha' + \beta' + \gamma = 2\gamma \tag{I.1.5}$$

and hence $\gamma > R$. Hence the triangle is obtuse, as to be shown.

By a similar reasoning, one shows that in case (ii), the assumption $OC > OA$ implies that the angle γ is acute. □

Problem I.1.7. *Provide a drawing for case* (i). *Provide two drawings for case* (ii), *one where the triangle $\triangle ABC$ is acute, and, as a catch, a second one where the $\triangle ABC$ is obtuse, nevertheless. (The triangle is obtuse because of a different obtuse angle.)*

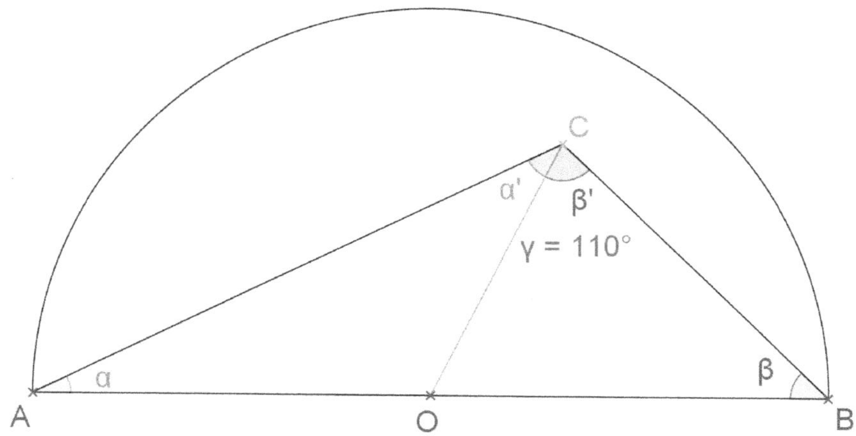

Figure I.1.8: A strengthening of Thales' Theorem, with vertex C inside the circle

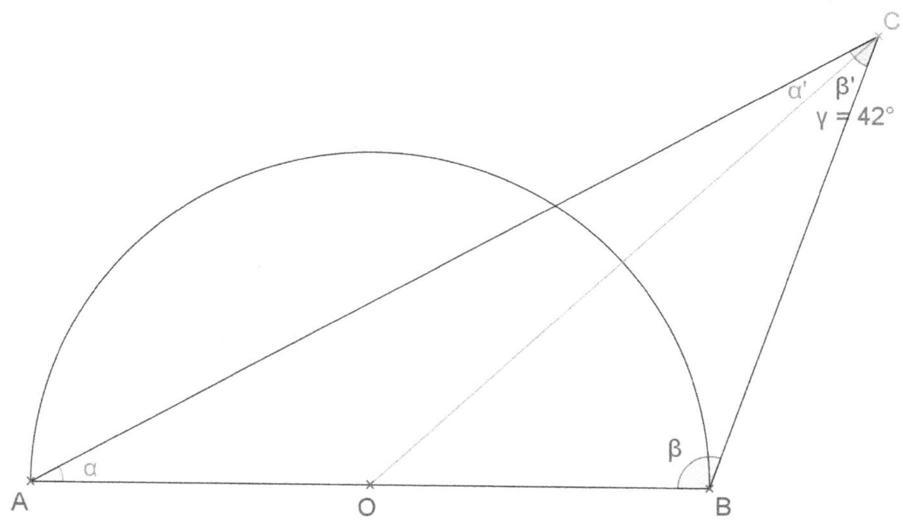

Figure I.1.9: A strengthening of Thales' Theorem, with vertex C outside the circle

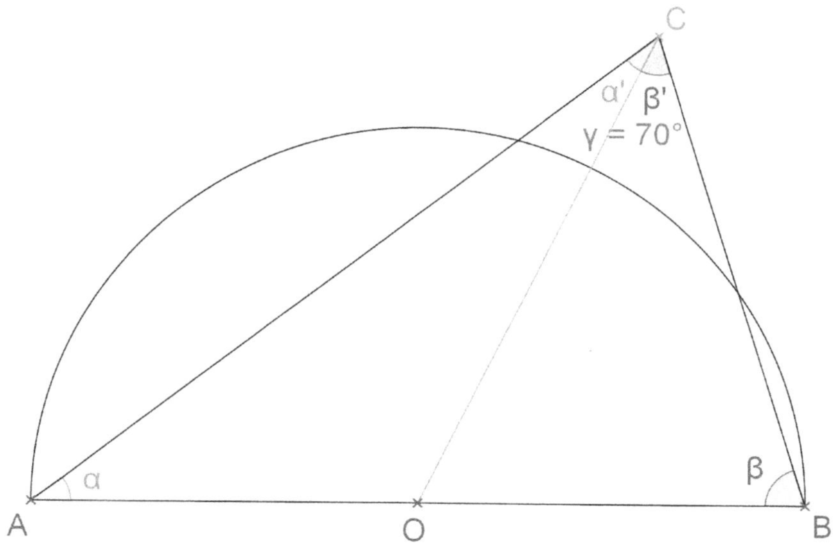

Figure I.1.10: A strengthening of Thales' Theorem, still another case with vertex C outside the circle

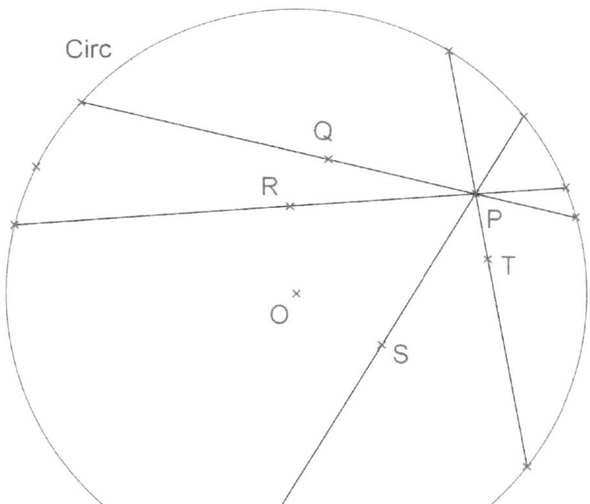

Figure I.1.11: Where lie the midpoints of all chords through point P?

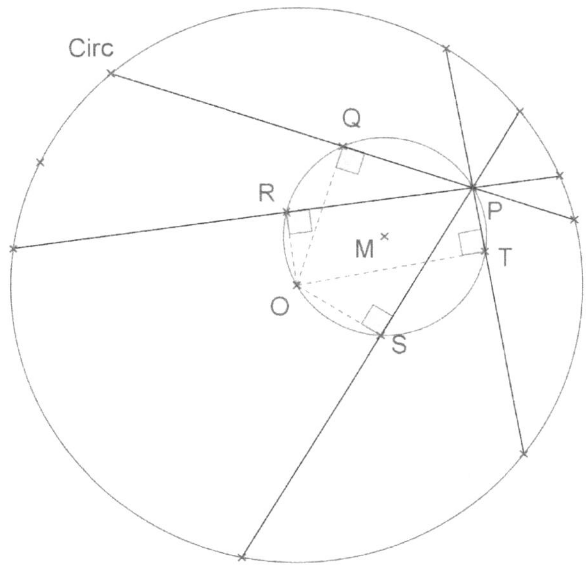

Figure I.1.12: The midpoints of chords lie on Thales' circle with diameter OP.

Problem I.1.8 (The midpoints of chords). *Given is a circle $\mathcal{C}$ with center O, and a point P inside $\mathcal{C}$. Describe the location of the midpoints of all chords through point P. Give a reason based on Thales' Theorem or its converse.*

Answer. The midpoints of all chords in the given circle $\mathcal{C}$ through the point P lie on a circle with diameter OP. Here is the reason: The midpoints are the foot points F of the perpendiculars dropped from the center O of the circle onto the respective chords. By the converse of Thales' Theorem, all the vertices F of the right angles with sides through the two points O and P lie on Thales' circle with diameter OP.

Corollary 3 (My strongest Thales' Theorem). *Given is any circle and any angle. Any two of the following statements* (i) (ii) (iii) *imply the third one.*

(i) *The two sides of the angle cut the circle at the endpoints of a diameter.*

(ii) *The vertex of the angle lies on the circle.*

(iii) *The angle is a right angle.*

Problem I.1.9. *Give explanations how to obtain a complete proof of the three parts of the strong Thales' Theorem 3, using all material from this section.*

Answer. The theorem has three parts:

"(i) and (ii) $\Rightarrow$ (iii)": This is simply the original Thales' Theorem I.1.1.

"(i) and (iii) $\Rightarrow$ (ii)": This is the Converse Thales' Theorem I.1.3.

"(ii) and (iii) $\Rightarrow$ (i)": This is consequence of Euclid III.20 (Theorem I.1.4) and Euclid III.22 (Theorem I.1.6). By Euclid III.20, the central angle $\angle AOB$ is twice the circumference angle with vertex <u>on</u> the long arc. By Euclid III.22, the circumference angles on the long and the short arc add up to two right angles.

Assume that A, B and C are any three points on a circle and $\angle ACB$ is a right angle. Let D be the other endpoint of diameter CD. By Euclid III.22, the angle $\angle ADB$ is a right angle, too.

Either C or D lies on the long arc AB, or AB is a diameter. Assume that C lies on the long arc AB. By Euclid III.20, we see that the central angle $\angle AOB$ is two right angles. If D lies on the long arc, we get the same conclusion that the central angle $\angle AOB$ is two right angles.. Hence AB is a diameter, anyway.

I.1.2 Rectangles and the Converse Thales Theorem

As already stated in the section on Legendre's Theorems, a *rectangle* is defined to be a quadrilateral with four right angles. The segments connecting the opposite vertices are called the *diagonals* of the rectangle. Obviously, a diagonal bisects a rectangle into two right triangles. They turn out to be congruent. Conversely, one can built a rectangle from two congruent right triangles. At first, we use this idea for an independent proof of the converse of Thales' Theorem 1. Secondly, the same idea helps to prove the rather obvious, but important properties of a rectangle.

Independent proof of the converse of Thales' Theorem 1. Given is the right triangle $\triangle ABC$, with the right angle at vertex C. The idea of this independent proof is constructing a rectangle from the right triangle.

Step 1: We transfer the angle $\alpha = \angle BAC$ to vertex B in order to produce the congruent angle $\angle ABD = \alpha$ on the side of hypothenuse AB opposite to vertex C. Furthermore, we transfer the segment AC to obtain the congruent segment $BD \cong AC$ on the newly produced ray.

Question. Show the congruence

$$\triangle ABC \cong \triangle BAD \qquad \text{(up-down)}$$

Which congruence theorem do you use?

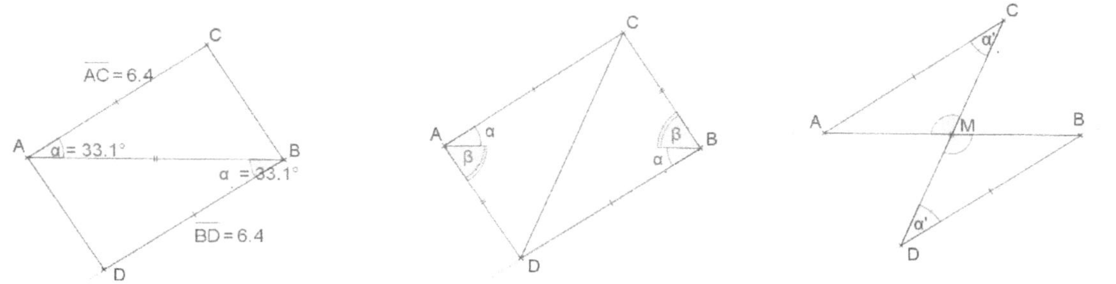

Figure I.1.13: (up-down) $\triangle ABC \cong \triangle BAD$ (left-right) $\triangle CAD \cong \triangle DBC$ (scissors) $\triangle AMC \cong \triangle BMD$

Answer. This follows by SAS congruence. Indeed, the two triangles have the common side AB, the two sides $AC \cong BD$ are congruent, and the angles $\angle BAC \cong \angle ABD$ are congruent, both by construction.

Step 2: The quadrilateral $\square ADBC$ obtained from the two triangles has remarkable symmetry. It is now bisected along its other diagonal CD to obtain a second pair of congruent triangles

(left-right) $$\triangle CAD \cong \triangle DBC$$

Question. Explain how this congruence is shown.

Answer. This follows by SAS congruence. Because of the first congruence (up-down), we have two pairs of congruent sides: $AC \cong BD$ by construction, and $BC \cong AD$ as a consequence of step 1. Furthermore, there are two pairs of congruent z-angles: the angles $\angle BAC \cong \angle ABD = \alpha$ are congruent by construction. The angles $\angle ABC \cong \angle BAD = \beta$ are congruent as a consequence of (up-down). Hence angle addition yields $\angle CAD \cong \angle DBC$. Finally, we get the claim (left-right) via SAS congruence.

The congruence (left-right) now yields two further pairs of congruent z-angles:

$$\angle ACD \cong \angle BDC = \alpha' \quad \text{and} \quad \angle CDA \cong \angle DCB = \beta'$$

Step 3: Since points C and D lie on different sides of line AB, the segment CD intersects this line at midpoint M.

Question. Use the congruence

(scissors) $$\triangle AMC \cong \triangle BMD$$

in order to show that the diagonals AB and CD bisect each other at their common midpoint M. By which theorem is this congruence proved?

Answer. The congruence (scissors) is obtained with the SAA congruence theorem. To use this theorem, we need the congruent sides $AC \cong BD$, the adjacent pair of congruent z-angles α', and the vertical angles at vertex M. Hence one gets the congruent triangles (scissors). Finally, we get the segment congruences

$$AM \cong BM = m' \quad \text{and} \quad CM \cong DM = m$$

Clearly point M lies between C and D, because these two points are on different sides of line AB by construction. But it is quite hard to confirm that M lies between A and B. Indeed, the congruence $AM \cong BM$ implies that point M lies between A and B.

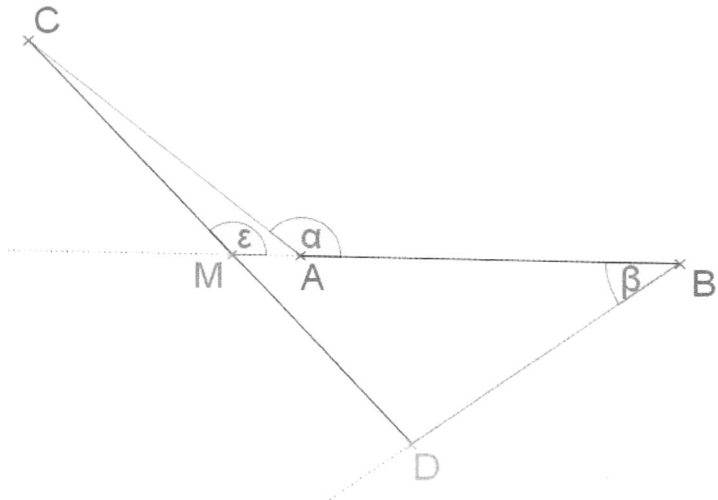

Figure I.1.14: Point A lying between M and B is impossible

Remark. See also figure on page 16 which is taken from the millenium edition of Hilbert's *"Grundlagen der Geometrie"*[10], page 26.

Step 4: We now get to the part of the reasoning which is valid only in Euclidean geometry. Because the angle sum in triangle $\triangle ABC$ is two right, and the angle at vertex C is assumed to be a right angle, we conclude that $\alpha + \beta = R$. Similarly, one confirms that the quadrilateral $\square ADBC$ has a right angle at vertex B. Hence it is a rectangle.

The right angle at vertex A yields still another pair of congruent triangles

(overlap) $$\triangle ABC \cong \triangle DCB$$

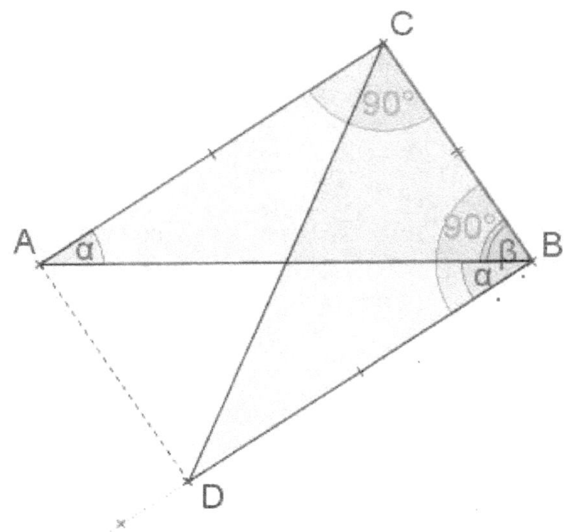

Figure I.1.15: The congruence of $\triangle BCA$ to $\triangle CBD$ implies that the diagonals of a rectangle are congruent

Question. Which theorem do you use to confirm this claim?

Answer. This follows by SAS congruence. Indeed, the two triangles have the common side $BC = CB$, the two sides $AC \cong DB$ are congruent by construction, and the angles $\angle BCA$ and $\angle DBC$ are both right. Indeed, $\angle BCA$ was assumed to be a right angle, whereas

$$\angle DBC = \angle DBA + \angle ABC \cong \alpha + \beta = R$$

because of congruent z-angles α by construction, and the angle sum.

As a consequence of the congruence (overlap), we get $AB \cong CD$. Thus we have shown that the diagonals of the quadrilateral $\square ADBC$ are congruent.

Step 5: This is the final step to get the converse Thales Theorem. Because the diagonals

$$AB \cong CD$$

are congruent, and point M bisects them both:

$$AM \cong MB \quad \text{and} \quad CM \cong MD$$

Hence all four segments from point M to the vertices A, B, C and D are congruent. Hence vertex C lies on a semicircle with diameter AB, as to be shown. $\square$

Remark. As a variant for the final step 5, we compare the distance $AM \cong BM = m'$ with $CM \cong DM = m$.

Assume that $m' > m$. Across the longer side of a triangle lies the larger angle. Using this fact for triangle $\triangle AMC$, we conclude that $\alpha' > \alpha$. Similarly, using triangle $\triangle BMC$, we get $\beta' > \beta$. Hence by addition, we get $R = \alpha' + \beta' > \alpha + \beta$. Hence the angle sum in the right triangle $\triangle ABC$ would be less than two right angles, which is false in Euclidean geometry.

Similarly, the assumption $m' < m$ leads to a contradiction, too. The only remaining possibility is $m = m'$. Hence vertex C lies on a semicircle with diameter AB and center M, as to be shown.

Problem I.1.10. *Which steps are still valid in hyperbolic geometry. Compare these quantities: m with m', α with α', and β with β'.*

Answer. Steps 1,2 and 3 are still valid. We still get $\alpha' + \beta' = 90°$, because the construction was started with a right triangle $\triangle ABC$ with right angle at vertex C. But, because the angle sum of a triangle is less than two right angles, one gets $\alpha + \beta < 90°$. Since $m < m'$ occurs if and only if $\alpha < \alpha'$ and $\beta < \beta'$, these inequalities are all three true. The figure on page 18 indicates the differing angles and segments using different colors. Remarkably enough, one can take the proof a step further

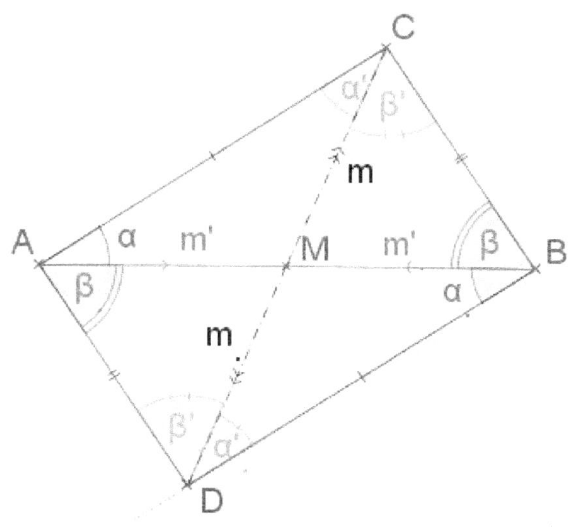

Figure I.1.16: What steps 1,2 and 3 yield in hyperbolic geometry

without appeal to Euclidean geometry: The perpendiculars dropped from M onto the four sides of the quadrilateral $\square ADBC$ lie on two lines, but these two lines are not perpendicular to each other.

We shall now recapitulate the same ideas, in order to derive the basic properties of a rectangle.

Corollary 4 (The basic properties of the rectangle). *Given is a rectangle. This is a quadrilateral, about which it is* only *assumed that it has four right angles.*

(i) *A diagonal partitions the rectangle into two congruent triangles.*

(ii) *The opposite sides of a rectangle are congruent.*

(iii) *The opposite sides of a rectangle are parallel.*

(iv) *The two diagonals of a rectangle bisect each other.*

(v) *The diagonals of a rectangle are congruent.*

(vi) *The diagonals of a rectangle intersect at the center of its circum circle.*

Proof. By definition, a rectangle is just a quadrilateral with four right angles—that is all what is assumed. The diagonal AB bisects the rectangle $\square ADBC$ into two right triangles $\triangle ABC$ and $\triangle ABD$. Again, let $\alpha = \angle CAB$ and $\beta = \angle ABC$ be the angles of the first of these triangles at vertices A and B, respectively.

We now use the fact that the angle sum in a triangle is two right angles. [1] Hence $\alpha + \beta + R = 2R$ and $\alpha + \beta = R$. The given rectangle has a right angle at vertex C, hence angle subtraction at vertex A now confirms that the second lower triangle $\triangle ADB$ has the $\angle BAD = R - \alpha = \beta$ at vertex A.

Thus we have obtained a pair of congruent z-angles $\beta = \angle ABC \cong \angle BAD$. Similarly, we get a second pair of congruent z-angles $\alpha = \angle BAC \cong \angle ABD$. With the ASA congruence theorem, we can now confirm the two triangles $\triangle ABC \cong \triangle BAD$ are congruent. Finally, the construction done above can be used once more to rebuild the rectangle from these two congruent right triangles.

We see that a diagonal partitions the rectangle into two congruent right triangles, and the opposite sides of the rectangle are congruent. Too, they are parallel because of the congruent z-angles.

The two diagonals of a rectangle are congruent because of

(overlap) $$\triangle BCA \cong \triangle DAC$$

[1] Actually, while proving the second Legendre Theorem, we have shown that the existence of rectangle already implies this fact about the angle sum.

Let M be the intersection point of the diagonals. The congruence (overlap) implies that the triangle $\triangle AMC$ has congruent base angles α. Hence the converse isosceles theorem implies $MA \cong MC$. Similarly, one can show that the triangle $\triangle BMC$ has congruent base angles β and the converse isosceles theorem implies $MC \cong MB$. Hence the two diagonals of a rectangle bisect each other. and their intersection point is the center of its circum circle. □

Problem I.1.11. *In Euclidean geometry, one defines a parallelogram to be a quadrilateral, the opposite sides of which are parallel. Which parts of the Corollary 4 are true for all parallelograms, in Euclidean geometry.*

Answer. Items (i) true (iv) are true for any parallelogram, but items (v) and (vi) are not true in general.

I.1.3 Construction of Tangents to a Circle

Problem I.1.12 (Tangents to a circle). *Given is a circle $\mathcal{C}$, with center O and a point P outside of $\mathcal{C}$. Construct the tangents from point P to the circle $\mathcal{C}$. Actually do and describe the construction!*

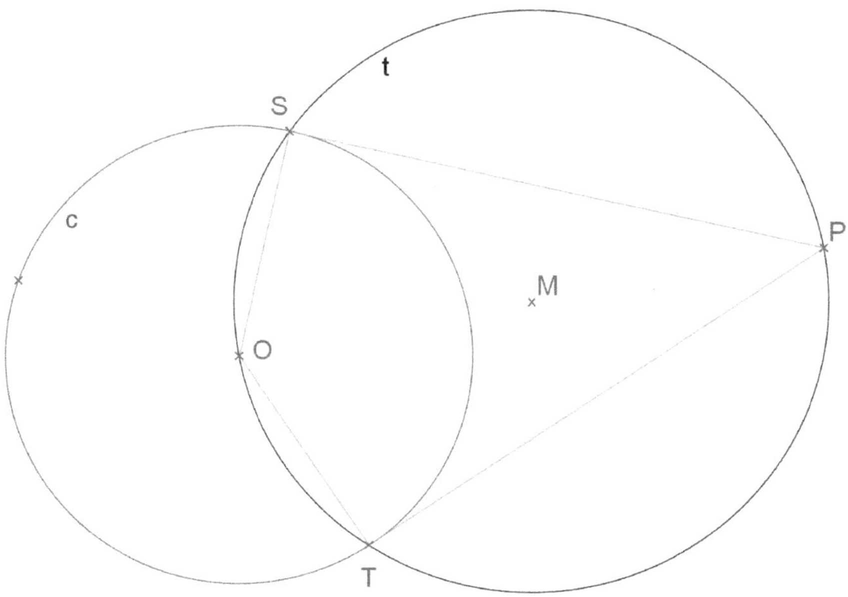

Figure I.1.17: Tangents to a circle

Construction I.1.1 (Tangents to a circle). *One begins by constructing a second circle $\mathcal{T}$ with diameter OP. (I call this circle the Thales circle over the segment OP). The Thales circle intersects the given circle in two points T and S. The lines PT and PS are the two tangents from P to circle $\mathcal{C}$.*

Validity of the Construction. By Thales theorem, the angle $\angle PTO$ is a right angle, because it is an angle in the semicircle over diameter OP. Since point T lies on the circle $\mathcal{C}$, too, the segment OT is a radius of that circle. By Euclid III. 16,

`The line perpendicular to a diameter is tangent to a circle.`

Since TP is perpendicular to the radius OT, and hence to a diameter, it is a tangent of circle $\mathcal{C}$. □

I.1.4 A Bid of Philosophy

If a mathematician learns such a nice theorem, as Thales' theorem is for sure, what does he want to do with it? [2] How does mathematics and other clever people benefit from it, after putting all issues of priority aside? Here are some general considerations:

Generalize the theorem. This can mean either getting along with less assumptions, or just putting the given assumptions into a more general context.

Simplify the statement of the theorem.

Sharpen the conclusions. Expressing the conclusions in another way and simplifying the statement can both help to sharpen them.

Ask whether a converse holds. There can be more than one version for the converse, depending one how the theorem is formalized. Too, in case the converse fails to be true, one can ask for similar statements of which the converse does hold.

Consider special cases. They can look surprisingly different—for example, some assumption may turn out to be too obvious to be stated explicitly.

Find applications. Are there constructions or algorithms which follow from the theorem.

[2] Well, in case he has discovered the theorem himself, one should publish it—with the possibility in mind that somebody else has already discovered something similar before.

Simplify the proof of the theorem.

Build a theory. Find the natural place for the theorem in a larger context.

Problem I.1.13. *We want to construct the tangents through a given point outside to a given circle. Which one of construction* (I.1.1) *and construction* (V.12.1) *— explained in the section on neutral geometry of circles and continuity—remains valid in hyperbolic geometry? Explain why.*

Answer. Construction (I.1.1) is no longer valid in hyperbolic geometry. On the other hand, construction (V.12.1) remains valid in hyperbolic geometry.

Construction (I.1.1) is not valid in hyperbolic geometry, because the angle sum of a triangle is less than two right, and hence Thales' theorem does not hold in hyperbolic geometry.

On the other hand, construction (V.12.1) depends only on SAS-congruence, and the fact that tangent and radius being perpendicular. These statements hold in hyperbolic geometry, too.

Problem I.1.14. *Once more: given is a circle and a point outside of it. We have to construct the tangent from the point to the circle.*

One could simply put the straightedge through this point and move it around until its line just touches the circle. What can one say on the one side in favor, and what on the other side against this "lazy-boy" construction? Gather all arguments coming to your mind.

Answer. Arguments in favor. The "lazy-boy construction" is quick and easy. The accuracy with which the tangent is obtained is as good as for either one of the traditional constructions—as long as one wants to obtain only the tangent and not the touching point.

<u>After the fact</u>—once it is known that the construction of the tangent is possible with the traditional tools—it turns out that the "lazy-boy construction" does not allow constructions for other problems than those solvable with the traditional tools. The "lazy-boy construction" is just an additional tool helping to do things simpler, in the same sense as the geometrical triangle does that.

Too, the "lazy-boy construction" would still work in hyperbolic geometry. □

Arguments against the easy construction. The "lazy-boy construction" does not allow to obtain the touching point with any reasonable accuracy. One can of course find the touching point by an additional step—dropping the perpendicular from the center of the circle onto the tangent. But in that way, the number of steps needed is nearly the same as for the traditional construction of the tangent.

Before establishing the fact that construction of the tangent is possible with the traditional tools, the only way to formally justify this procedure is by requiring a new axiom— on the same footing as it has been suggested for origamy, or the two marked ruler, or any other extended tools of construction.

Question. Is the axiom

> "The tangent from a point to a circle can be obtained by appropriate placement of the straightedge (lazy-boy construction)."

equivalent to the circle-line intersection property, or to the circle-circle intersection property? Or is it a weaker axiom?

Question. How often does one really want new axioms?

Because of such questions, the "lazy-boy construction" cannot simply replace the traditional construction, nor does it make the traditional construction and its proof superfluous. □

My last word. In the end, I think one can use this simple way to obtain the tangent, if it is not necessary to have the touching point. □

I.1.5 Common Tangents of Two Circles

Problem I.1.15 (Common tangents of two circles). *How many common tangents do two circles have. Informally draw all different cases, with* $0, 1, 2, 3, 4$ *common tangents. Describe how they arise.*

Answer. For any two different circles, there are five possibilities regarding their common tangents:

(0) One circle lies inside the other. They have no common tangents.

(1) One circle touches the other from inside. There is one common tangent, located at this touching point.

(2) The two circles intersect in two points. They have two common tangents, which lie symmetrically to the axis connecting the two centers.

(3) The two circles touch each other from outside. They have three common tangents.

(4) The two circles lie outside of each other. They have four common tangents. These are two pairs lying symmetrically to the axis connecting the two centers.

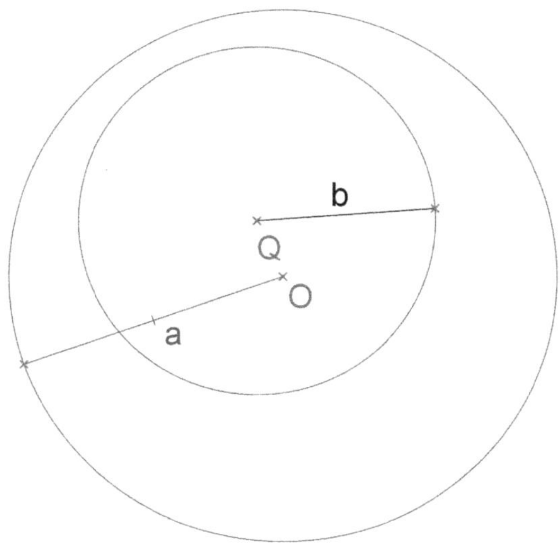

Figure I.1.18: Two circles with no common tangents.

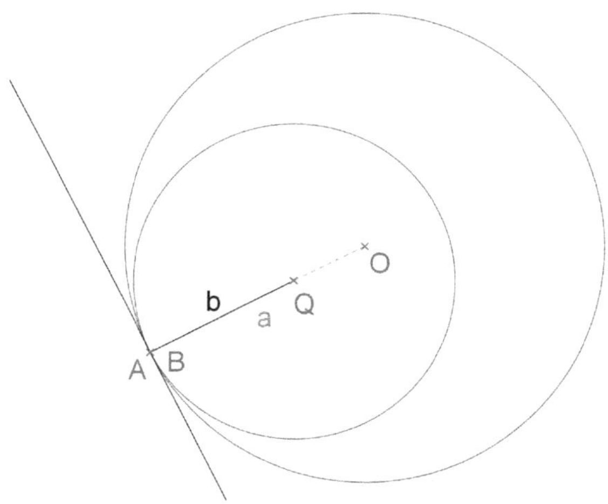

Figure I.1.19: Two circles with one common tangent.

Remark. Let the circles have center O and radius a, and center Q and radius b. These cases correspond to:

(0) $|OQ| < |b - a|$: One circle lies inside the other.

(1) $0 < |OQ| = |b - a|$: One circle touches the other from inside.

(2) $|b - a| < |OQ| < a + b$: The two circles intersect in two points.

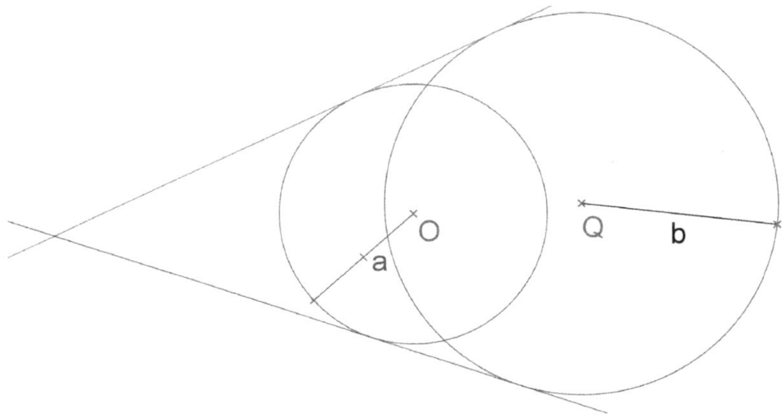

Figure I.1.20: Two circles with two common tangents.

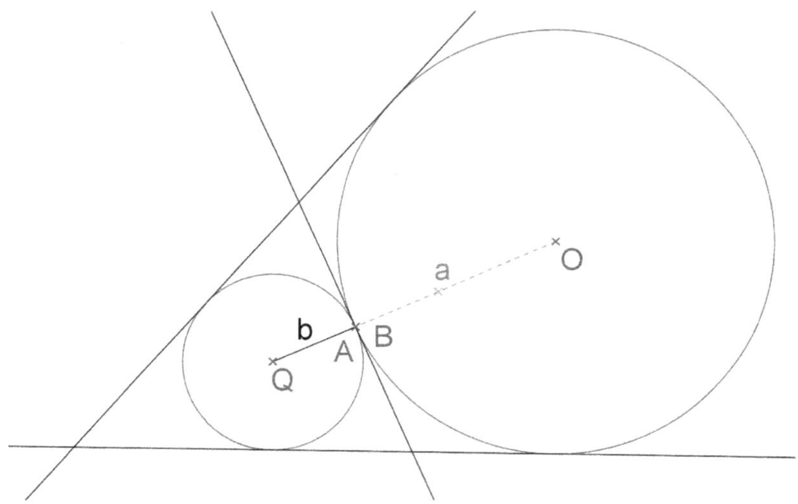

Figure I.1.21: Two circles with three common tangents.

(3) $|b-a| < |OQ| = a+b$: The two circles touch each other from outside.

(4) $a+b < |OQ|$: The two circles lie outside of each other.

(∞) $0 = |OQ| = |b-a|$: The two circles are equal to each other.

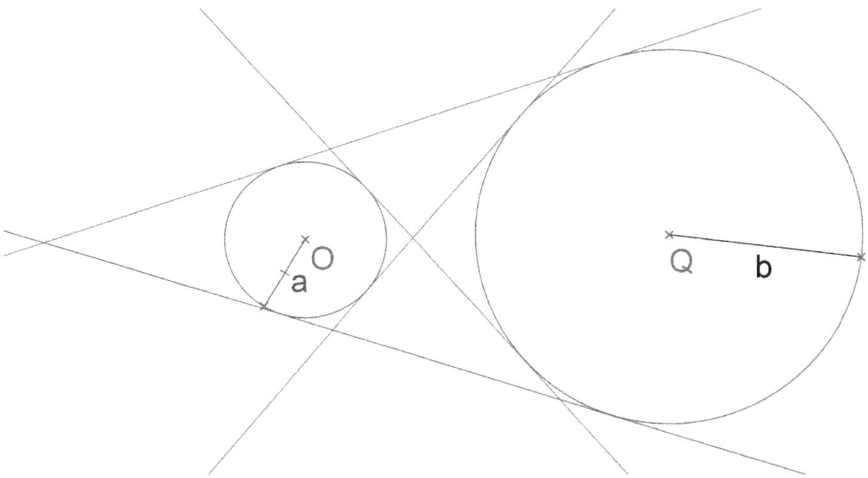

Figure I.1.22: Two circles with four common tangents.

Problem I.1.16 (Construction of the common tangents of two circles). *Given are two circles, with center O and radius a, and center Q and radius b. They lie outside of each other. Explain how to construct, with Euclidean tools, the common tangents of the two circles. Actually do and describe the construction!*

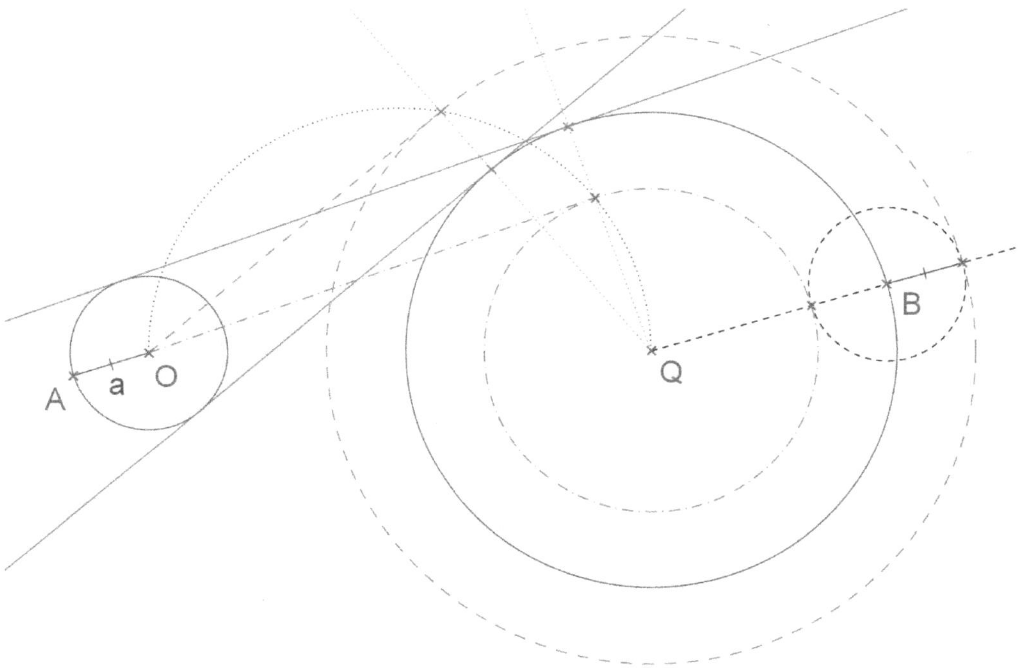

Figure I.1.23: Construction of two of the four common tangents for two circles lying outside each other.

The drawing on page 27 constructs just two common tangents, using construction I.1.2. The other two are their mirror images by line OQ connecting the centers.

Construction I.1.2 (Common tangents of two circles). *One constructs two new circles around one of the given centers, say Q, choosing as their radii the sum $a+b$ and difference $|a-b|$ of the radii a and b of two given circles. Next one constructs the tangents from the second center O to these two new circles. Too, one needs the radii where these tangents touch. The common tangents are produced by parallel shifts of these tangents, to both sides, by a distance given by the radius of the original circle around O.*

Shifting the tangents to the circle of radius $a+b$ yields the inner common tangents, whereas shifting the tangents to the circle of radius $|a-b|$ results in the outer common tangents.

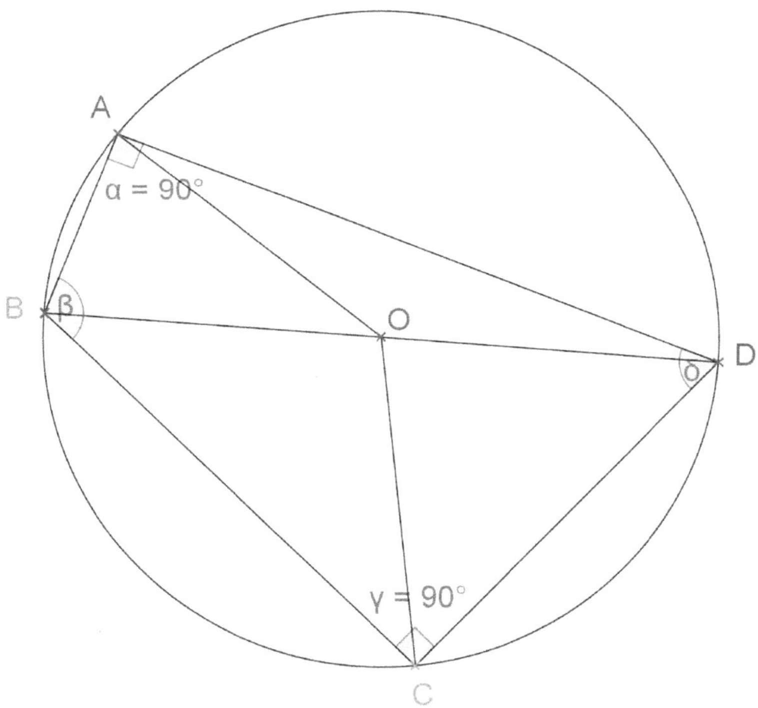

Figure I.1.27: A special quadrilateral

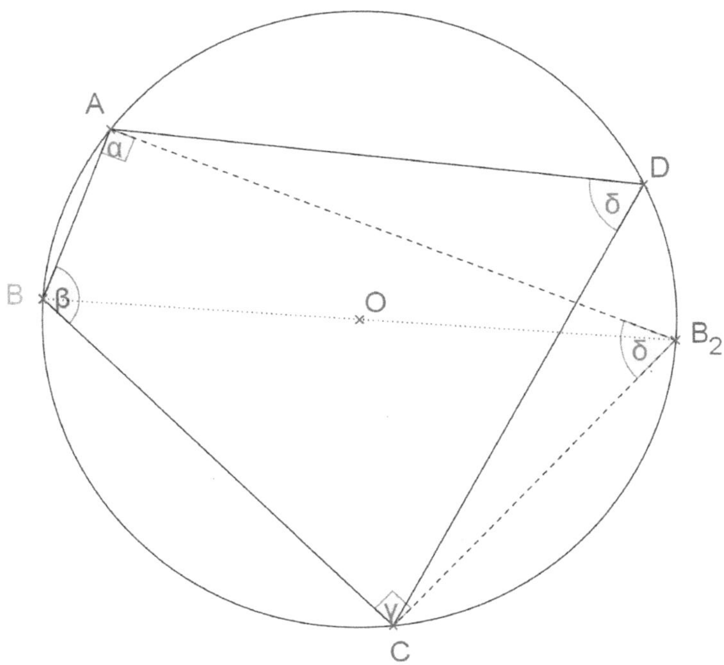

Figure I.1.28: Opposite angles in a quadrilateral with a circum circle

the angle sums of the two triangles. Since the angle sum in a triangle is two right angles, the angle sum of a quadrilateral is $4R$.

By Thales' theorem, these are both right triangles, with right angles at vertices A and C. After subtraction of the two right angles at vertices A and C, the sum of the two remaining angles of the quadrilateral at vertices B and D is $2R$. □

Theorem I.1.7 (Euclid III.22). *The opposite angles of a circular quadrilateral add up to two right angles.*

Proof. In the circum circle, either one of the arcs ABC or ADC is less or equal the other one, and hence less or equal a semicircle. We can assume that arc ABC is less or equal a semicircle. Let B_2 be the second endpoint of diameter BB_2. Now B_2 lies on the arc ADC. As shown in the last problem, the angles at opposite vertices B and B_2 add up to two right angles in the quadrilateral $\square ABCB_2$

$$\angle ABC + \angle AB_2C = 2R \tag{I.1.7}$$

By Euclid III.21, the circumference angle of arc ABC is

$$\angle ADC \cong \angle AB_2C \tag{I.1.8}$$

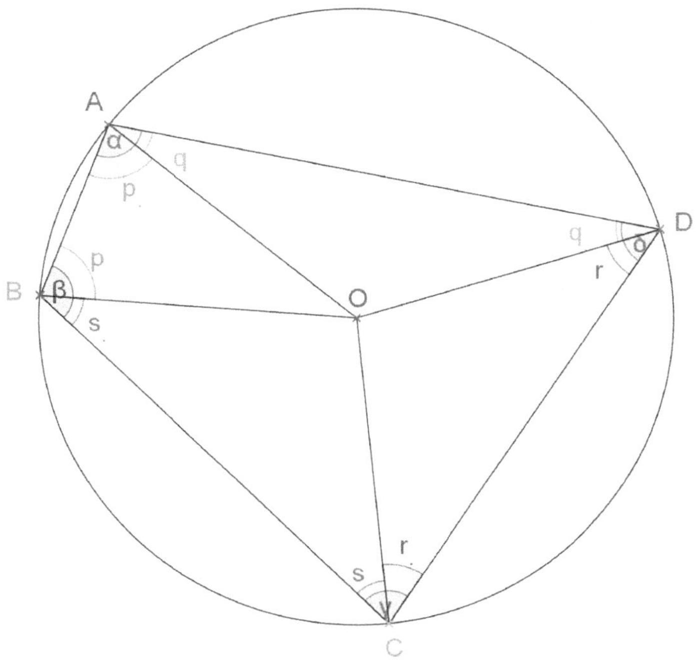

Figure I.1.29: Use isosceles triangles

Together, we conclude

$$\angle ABC + \angle ADC \cong \angle ABC + \angle AB_2C = 2R \tag{I.1.9}$$

□

Remark. Here is an argument to get Euclid III.22 at once. We draw the four radial segments OA, OB, OC and OD and partition the quadrilateral $\square ABCD$ into four isosceles triangles. Let p, q, r and s be their respective base angles.

Because of angle addition at each vertex, the sum of the two angles of the quadrilateral $\square ABCD$ at the opposite vertices A and C is

$$\alpha + \gamma = (p + q) + (r + s) = (p + s) + (q + r) = \beta + \delta \tag{I.1.10}$$

Since the sum of angles at all four vertices is $\alpha + \beta + \gamma + \delta = 4R$, we conclude

$$\beta + \delta = \alpha + \gamma = 2R \tag{I.1.11}$$

Proposition I.1.1 (Another equivalent formulation of Euclid III.22). *An interior angle of a circular quadrilateral is congruent to the exterior angle at the opposite vertex.*

Proposition I.1.2 (Strengthening of Euclid III.22). *Assume vertices B and D lie on different sides of AC. The opposite angles $\beta + \delta$ of a quadrilateral $\square ABCD$ add up to*

(i) *more than two right angles if and only if vertex D lies inside the circum circle of triangle ABC,*

(ii) *two right angles if and only if vertex D lies on the circum circle,*

(iii) *less than two right angles if and only if vertex D lies outside the circum circle of triangle ABC.*

Problem I.1.19. *Draw a convex quadrilateral $\square ABCD$. The vertices A, B, C, D are named in alphabetic order as they follow each other around the quadrilateral. Draw any instance for these two examples:*

(a) *the opposite angles have sum $\alpha + \gamma = 120°$,*

(b) *the opposite angles have sum $\alpha + \gamma = 270°$.*

Draw a circle through A, B and C and explain how point D is located relative to the circle.

Answer. The angle sum of any quadrilateral is $4R$, and hence $\beta + \delta = 360° - \alpha - \gamma$.

In case (a), one gets $\beta + \delta = 360° - 120° = 240° > 180°$. Hence vertex D lies inside the circum circle of triangle ABC.

In case (b), one gets $\beta + \delta = 360° - 270° = 90° < 180°$, and hence vertex D lies outside the circum circle of triangle ABC.

Given is a circle with center O. Let AB be the chord and T be a point on the perpendicular to the radius at point A. We assume that center O does not lie in the same half plane of the chord AB as point T.

Proposition I.1.3. *The angle $\angle ABT$ is congruent to the circumference angle $\angle ACB$ with vertex C on the long arc.*

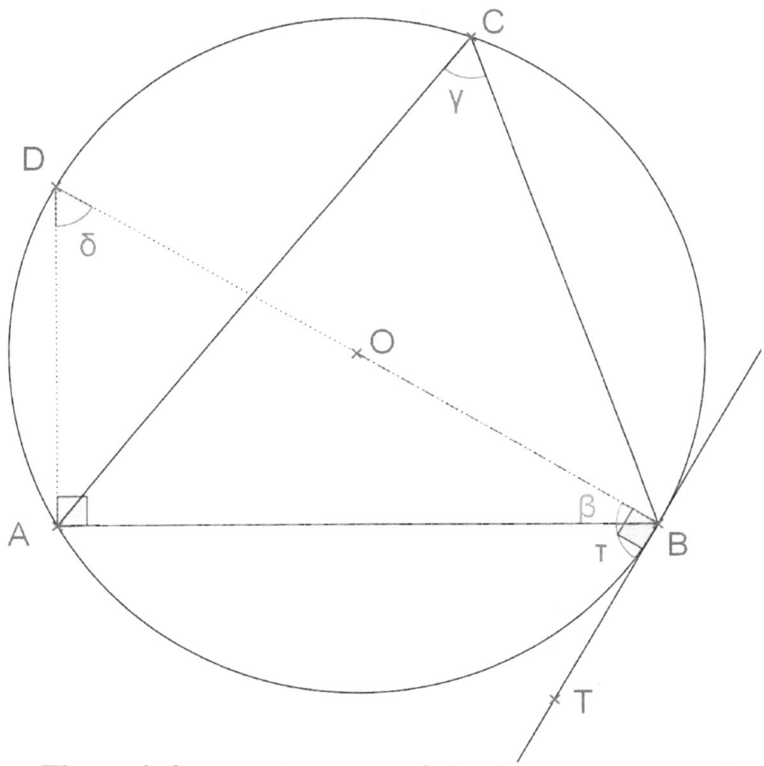

Figure I.1.30: The angle between tangent and chord is congruent to the circumference angle

Proof. Since we assume that center O does not lie in the same half planen of the chord AB as point T, the angle $\tau = \angle ABT$ is acute or right. We prove the statement first for the circumference angle of an appropriately chosen point D. Let BD be a diameter of the circle. By Thales' theorem, triangle $\triangle ABD$ is a right triangle. Because the angle sum of a triangle is $2R$, its two angles at vertices D and B, add up to a right angle:

$$\beta + \delta = R$$

On the other hand, angle addition at vertex B yields,

$$\beta + \tau = R$$

From these two equation and angle subtraction, we conclude that $\tau = \delta$. Thus the angle $\tau = \angle TAD$ between the perpendicular of the radius at point B and the chord AB is congruent to the circumference angle $\delta = \angle ADB$ of that chord. By Euclid III.21:

```
Two angles from points of the circle subtending the same arc
   are congruent.
```

Hence it does not matter that we have chosen a special position for point D. The circumference angle $\gamma = \delta = \tau$ for any point C on the long arc of AB is congruent to the angle τ between the chord and the tangent. □

I.1.7 On the Nature of the Tangent

Euclid III.16 contains the assertion that the line erected perpendicularly onto a diameter at its end is tangent to the circle. The fact that the tangent is orthogonal to a diameter is stated repeatedly in Euclid III.16, III.18 and III.19—without further clarification how the tangent is defined. Euclid formulates our result proposition I.1.3 in his proposition Euclid III.32:

```
The angle between a tangent line and a chord is congruent
   to the circumference angle of the arc corresponding to the
   chord.
```

I find this state of affair unsatisfactory. Hence I have <u>defined</u> the tangent to a circle as perpendicular to the radius (see definition V.12.6). Proposition V.12.19 states that a line is tangent to a circle if and only if it has exactly one point in common with the circle. This is just the second obvious property of the tangent, which does not involve limits. We now agree to use the notion of limits. We define the tangent as

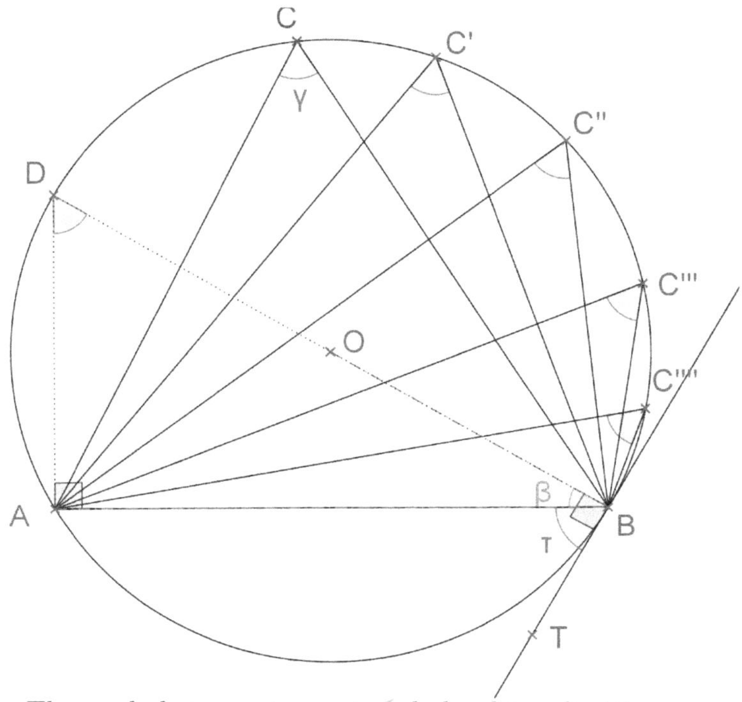

Figure I.1.31: The angle between tangent and chord is a limiting case of the circumference angle

the limiting position of a secant line, the two endpoints of which move together at the touching point of the tangent. The circumference angle of the fixed chord AB constantly stays γ. In the limit $C \to B$, one side of the circumference angle $\angle ACB$ becomes the chord AB, and the other side CB becomes the tangent t to the circle at point B. Hence, in the limit we see that the angle between the chord AB and the tangent t at its endpoint B still has the same value γ.

Proposition I.1.4. *We obtain the tangent as limiting position of a secant line, as the length of the secant approaches zero. In this setting, the angle between any chord and the tangent ray at its endpoint, pointing to the side of the short arc, is congruent to the circumference angle with vertex on the long arc.*

Corollary 8 (The tangent as limiting position of a short secant). *The tangent to a circle, defined as the limiting position of a small secant, is perpendicular to the radius at the touching point.*

Proof. Let t be the perpendicular to the radius OB at the touching point B. Let $\tilde{t}$

be the tangent obtained as the limiting position of a small secant with one endpoint B.

By proposition I.1.3, the acute angle between chord AB and t is the circumference angle γ. On the other hand, the remark above and proposition I.1.4 show that the acute angle $\tilde{t}$ between chord AB and tangent as limit is congruent to the circumference angle γ, too. Hence, by the uniqueness of angle transfer, we conclude $t = \tilde{t}$. In other words, the tangent is both the limiting position of a small secant, and the perpendicular to the radius at the touching point. □

Remark. Indeed, we have already obtained the same result even in neutral geometry. The most complete result is indeed Theorem V.12.3.

Remark. Historically, the exact nature of the tangent is not discussed at this point, or any point, by Euclid. Euclid III.16 contains the cryptic statement—I would say, the only cryptic statement in the entire work of Euclid:

> "The angle between the tangent line and the circle
> is less than any rectilinear angle."

I.1.8 The Two-Circle Lemma

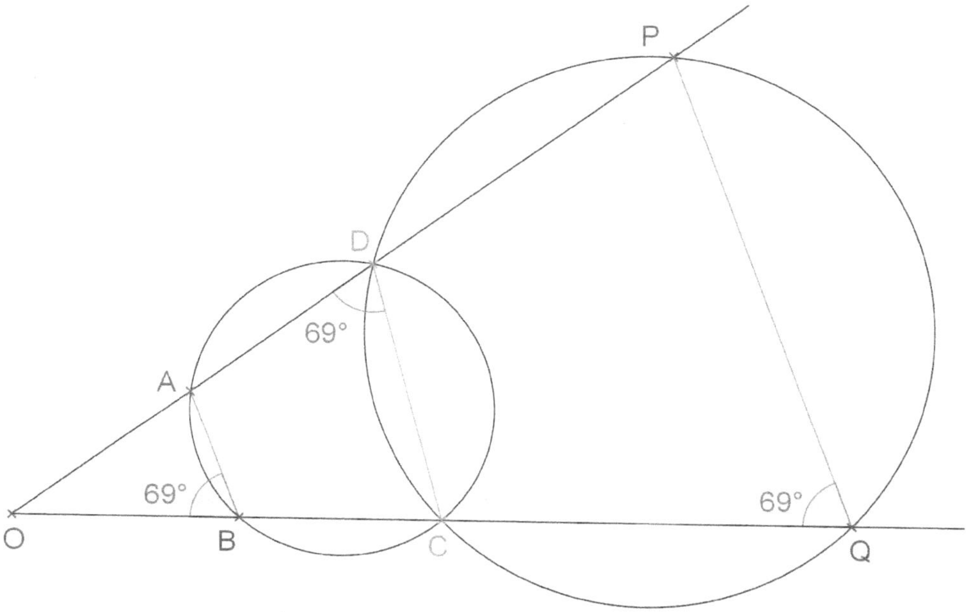

Figure I.1.32: The sides of an angle cutting two circles at the endpoints of their common chord cut them in two further parallel chords.

The following lemma is remarkable by itself, and shall be used again in the proof of Pappus' and Pascal's Theorems.

Lemma I.1.1 (Two Circle Lemma). *If the endpoints of the common chord of two circles lie on two lines, these lines cut the two circles in two further parallel chords.*

Problem I.1.20. *Explain the reason for the two circle lemma. To get a convenient picture, consider the situation drawn in the figure on page 42. The two given circles have the common chord CD. Let AB be a second chord in the left circle. Explain why the second chord PQ in the right circle is parallel to AB.*

Going over the proof, does it matter whether the two given lines are parallel or not?

Answer. We get a circular quadrilateral $\square ABCD$ inside the left circle $\mathcal{C}$, and a second circular quadrilateral $\square CDPQ$ inside the right circle $\mathcal{D}$. As a consequence of Euclid III.22, the opposite angles of a circular quadrilateral sum up to two right angles.

We check that the segments AB and PQ form congruent angles with one of the given lines. We need to compare the two angles $\angle OBA$ and $\angle OPQ$.

Let O be any point of line BC and O' any point on line AD such that the points mentioned occur on the two lines in the order $O*B*C*Q$, and $O'*A*D*P$. The angle $\angle OBA$ is exterior angle in the circular quadrilateral $\square ABCD$. As a consequence of Euclid III.22, it is congruent to the opposite interior angle:

$$\angle OBA \cong \angle O'DC$$

Now the second angle is exterior angle in the second circular quadrilateral $\square CDPQ$. Hence it is congruent to the opposite interior angle:

$$\angle O'DC \cong \angle OQP$$

From these two congruences, we conclude $\angle OBA \cong \angle OQP$. Hence, by Euclid I.27, the lines AB and PQ are parallel. We see that it does not matter whether the two given lines intersect or not.

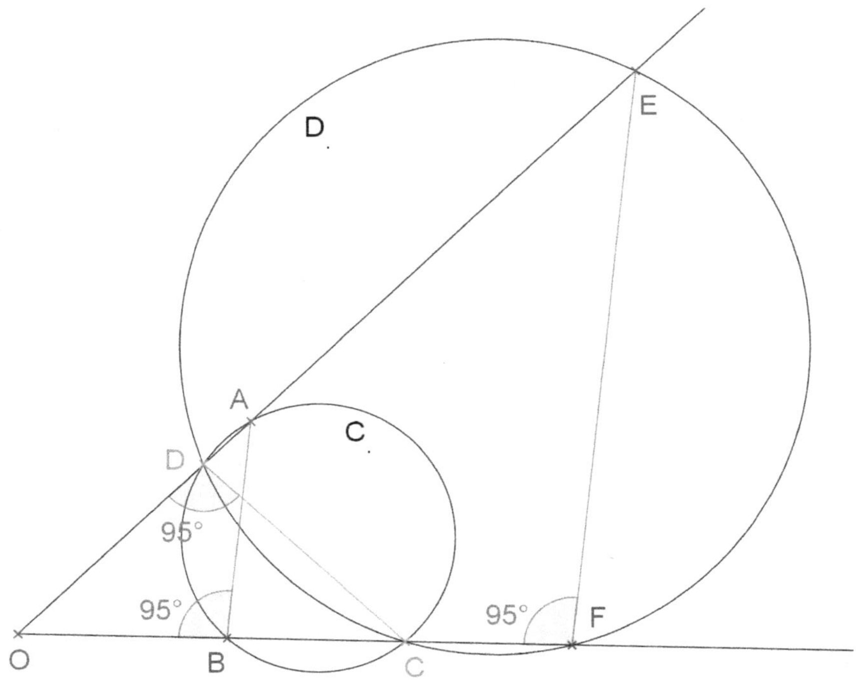

Figure I.1.33: Another example for the two-circle lemma.

Question. Explain how the proof has to be modified in the case drawn in the figure on page 43.

Answer. In this case, the circular quadrilateral has intersecting opposite sides AB and CD. In this case, the interior angles at opposite vertices are congruent. Hence, the exterior angles at opposite vertices are congruent, too.

I.2 Simple Euclidean Geometry

These exercises deal with some interesting questions from Euclidean geometry on the level of Euclid's classic. The parallel axiom and its consequences, and the circle intersection properties are assumed to hold.

I.2.1 Fundamental Constructions with Euclidean Tools

Definition I.2.1 (Traditional Euclidean tools). Constructions with *traditional Euclidean tools* are constructions using only straightedge and compass, done in a finite number of steps of the following types:

1. drawing a line through two different points
2. drawing a circle with a given center through a given point
3. intersecting two lines
4. intersecting a circle and a line
5. intersecting two circles
6. choosing an arbitrary point
7. choosing a point on a given line
8. choosing a point on a given circle
9. deciding whether two points are equal or different
10. deciding whether a point lies on a line
11. deciding whether a point lies on a circle

Remark. For the enumeration and count of a construction, we do not count producing the given pieces, neither pieces that were only drawn to facilitate the understanding. But producing the resulting pieces is counted.

Problem I.2.1. *Explain how an angle is bisected with traditional Euclidean tools. Enumerate and count the steps needed, according to definition V.12.10, and provide a drawing.*

Answer. Let $\angle(h, k)$ with vertex A be given.

1. We choose a point $B \neq A$ on the ray h.
2. Draw circle(A, B).
3. Find the intersection point $2 \cap k$.
4. Draw circle$(3, A)$.
5. Draw circle(B, A).
6. Find the intersection point $4 \cap 5$ in the interior of the given angle $\angle(h, k)$.
7. Draw the ray $\overrightarrow{A, 6}$. This is the angle bisector to be constructed.

We count that seven steps are needed for the construction.

Problem I.2.2. *Given is a line l and a point lying on the line. Explain how the perpendicular is erected with traditional Euclidean tools.*

Answer. We draw any circle around the given point P and let A and B be the intersection points of line l with this circle. Next we draw two circles, with center A through point B, and with center B through point A. They intersect in two points, let C be one of them. The line PC is the perpendicular to line l at point P to be constructed. We get an extra check of accuracy, since this line goes through the other intersection point of the two circles, too.

Problem I.2.3. *Given is a line l and a point P not lying on the line. Explain how to construct the reflection point P' across the line with traditional Euclidean tools, without using line-circle intersection as a means of construction.*

Answer. We choose any two different points $A \neq B$ on the given line l and draw the circles with center A through point B, and with center B through point A. They intersect in two points: the given point P and a point $P' \neq P$, which is the reflection of point P across the line l to be constructed.

Problem I.2.4. *Given is a line l and a point not lying on the line. Explain how the perpendicular is dropped with traditional Euclidean tools. Avoid line-circle intersection as a means of construction.*

Answer. We choose any two different points $A \neq B$ on the given line l and draw the circles with center A through point B, and with center B through point A. They intersect in two points: the given point P and the reflection point Q. The line PQ is the perpendicular to the line l through point P to be constructed.

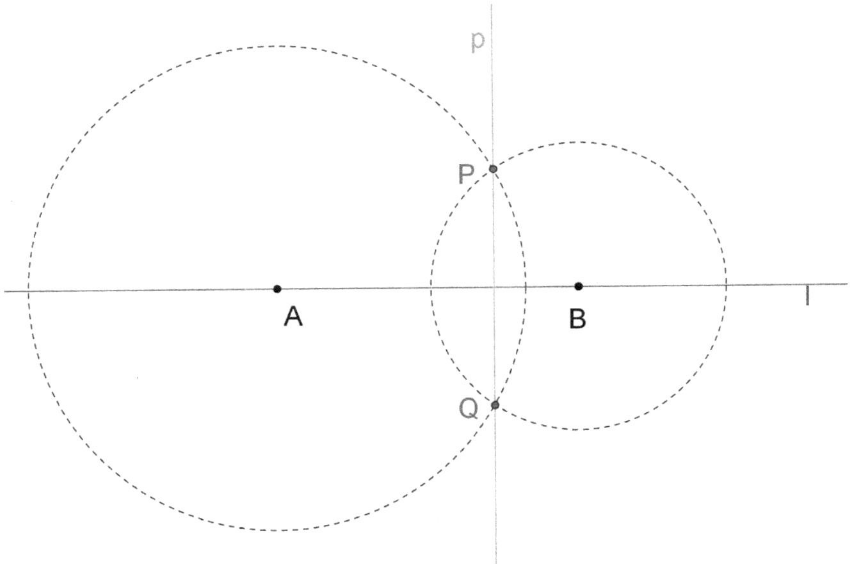

Figure I.2.1: Construction of the perpendicular with just two circles and their intersection.

Remark. The two points may lie on the same or different sides of the perpendicular, the construction works for any choice of the points $A \neq B$. The two circles may have different radius. In any case, they indeed intersect at two points.

Problem I.2.5. *Explain that two equilateral triangles the common side of which is any given segment, can be constructed with traditional Euclidean tools. How can the two triangles be distinguished.*

Answer. Let the segment AB be given. We draw the circles with center A through point B, and with center B through point A. They intersect in two points C and D. The triangles $\triangle ABC$ and $\triangle ABD$ are the two equilateral triangles with sides AB to be constructed. They lie on different sides of line AB.

Problem I.2.6. *Explain how the perpendicular bisector of a segment is constructed with traditional Euclidean tools.*

Answer. Let the segment AB be given. We draw the circles with center A through point B, and with center B through point A. They intersect in two points C and D. The line CD is the perpendicular bisector segment AB to be constructed.

Problem I.2.7. *Given is a line l, a point A not lying on the line, and a point B lying on the line. Construct a circle through points A and B that is tangent to the line. Provide a drawing and explain the major steps.*

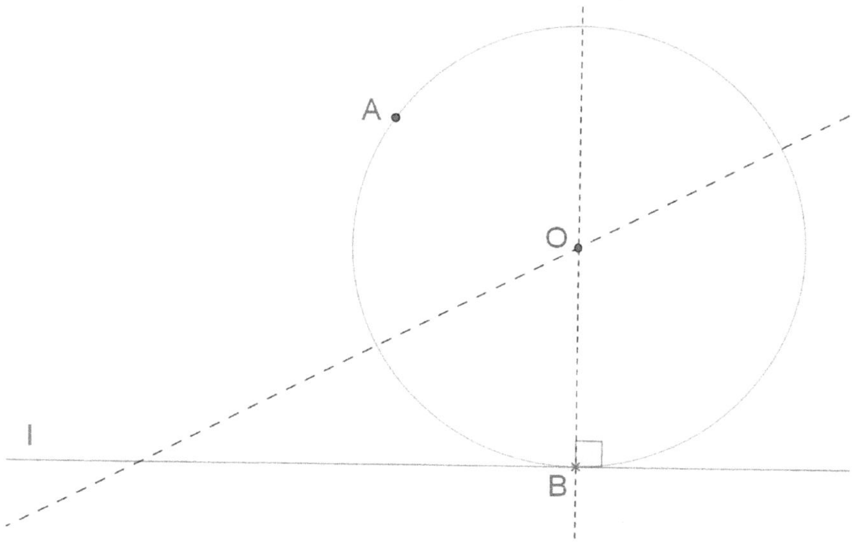

Figure I.2.2: Construction of the circle through A and B tangent to line l.

Answer. We erect the perpendicular at point B onto the given line l. Next we construct the perpendicular bisector of segment AB. The intersection of these two lines is the center O of the circle through A and B, tangent to line l, to be constructed.

I.2.2 Five Constructions of the Parallel

Given is a line l and a point P not lying on the line. We want to construct the parallel to line l through the point P.

Construction I.2.1 (The parallel as double perpendicular). *We draw a circle with center P through point A on the line l, and obtain a second intersection point B. We draw two further circles around A and B, both through point P. They intersect in a second point Q, too, which is the the reflection image of P across the line l. We draw the line $PQ \perp l$ and get the intersection points C and D with the first circle.*

Finally, we need the two circles with center C through point D, as with center D through point C. They intersect in two further points R and S. The line $RS \parallel l$ is the parallel to line l through point P.

Remark. As an extra check, we have obtained indeed three points P, R and S on the parallel.

Problem I.2.8. *Given is a line l and a point P not lying on the line. We want to construct the parallel to line l through the point P. Provide drawing with notations*

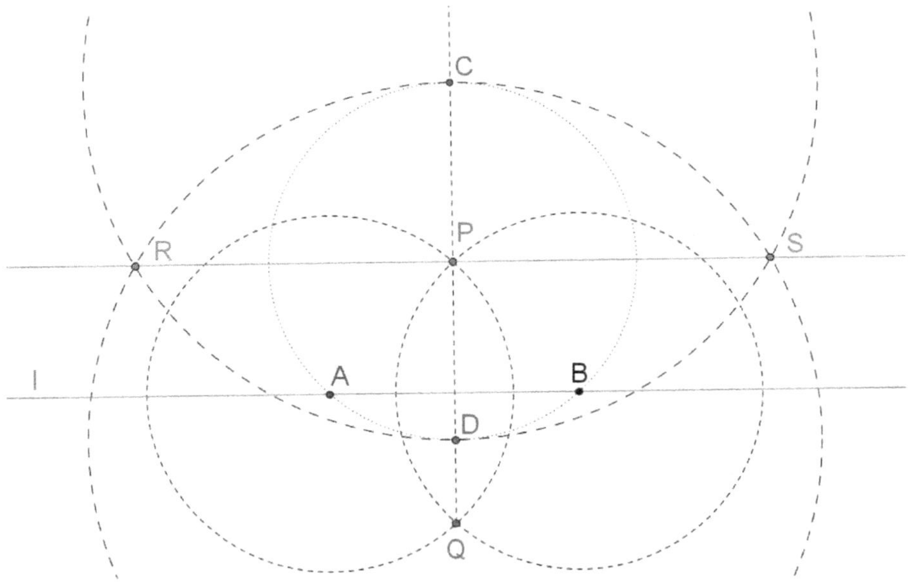

Figure I.2.3: The parallel as double perpendicular.

to illustrate the construction I.2.1 given above.

Remark. As an extra check, we can get the second endpoint F of the diameter BF of the first circle. The two circles with the centers F and A, both through point P. They intersect in a second point S. In this way, we get three points P, R and S on the parallel.

Construction I.2.2 (The parallel using Thales' circle). *We draw a circle with center P through point A on the line l, and obtain a second intersection point B. We draw a further circle around B through point P.*

Let point E be the other endpoint of the diameter AE of the first circle. Finally, we draw two circles with the centers E and B, both through point P. Finally, we draw two circles with the centers E and B, both through point P. They intersect in a second point R, too. The line $RP \parallel l$ is the parallel to line l through point P.

Justification of the construction. The angle $\angle ABE$ is a right angle by Thales' theorem since it is the angle in a semicircle. All three circle drawn have congruent radius $PA \cong PB \cong PE$. Hence the intersection points P and R of the circles around points B and E lie on the perpendicular bisector of segment BE. Hence lines BE and PR are perpendicular. Thus the line given line $l = AB$ and the line PR have a common perpendicular, and hence they are parallel. $\square$

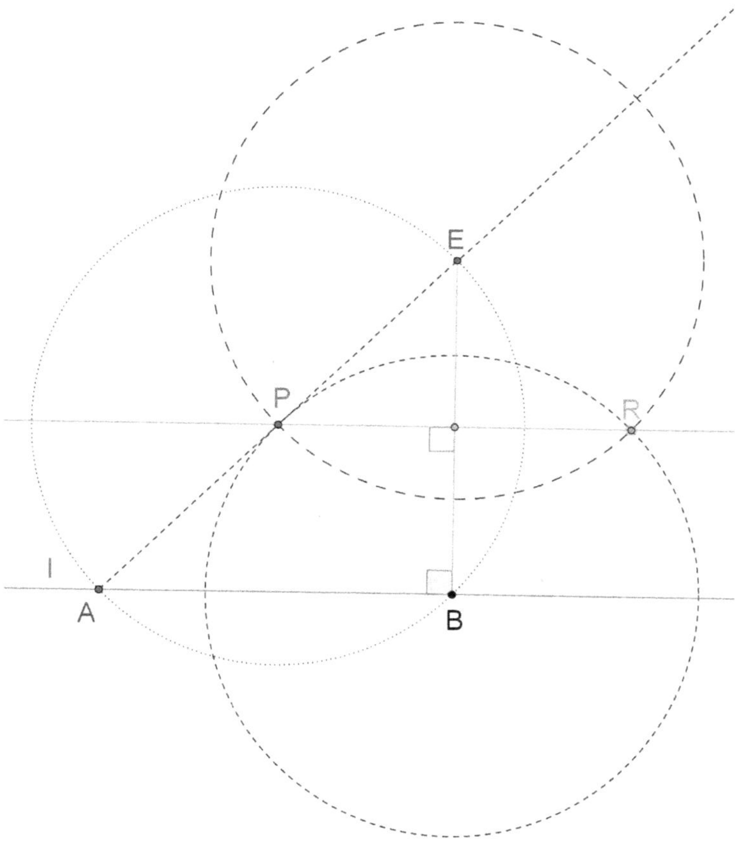

Figure I.2.4: The parallel using Thales' circle.

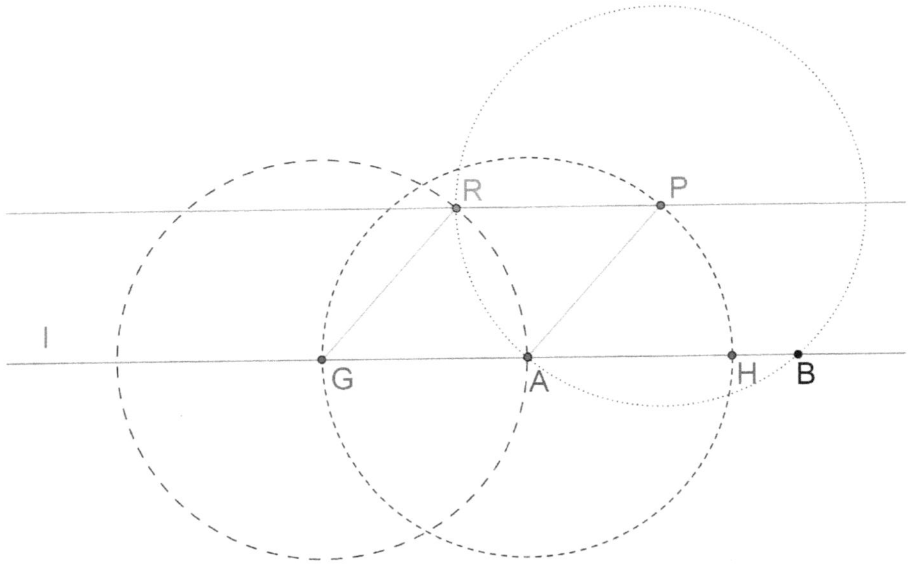

Figure I.2.5: The parallel obtained from a rhombus.

Construction I.2.3 (The parallel obtained from a rhombus). *We draw a circle with center P through point A on the line l. (We obtain a second intersection point B, too.)*

We draw the circle with center A through point P. It intersects the line l in the points G and H. We draw a third circle around G through A. It intersects the very first circle in point R. The line RP is parallel to the given line l since RP and GA are opposite sides of the rhombus $\square GAPR$.

Remark. We can draw a forth circle around H through A. It intersects the very first circle in point S. The line RP is parallel to the given line l since The segments SP and HA are opposite sides a rhombus and hence parallel. In this way, we get three points P, R and S on the parallel, which provides an extra check.

Problem I.2.9. *Given is a line l and a point P not lying on the line. We want to construct the parallel to line l through the point P. Provide drawing with notations to illustrate the construction I.2.3 given above.*

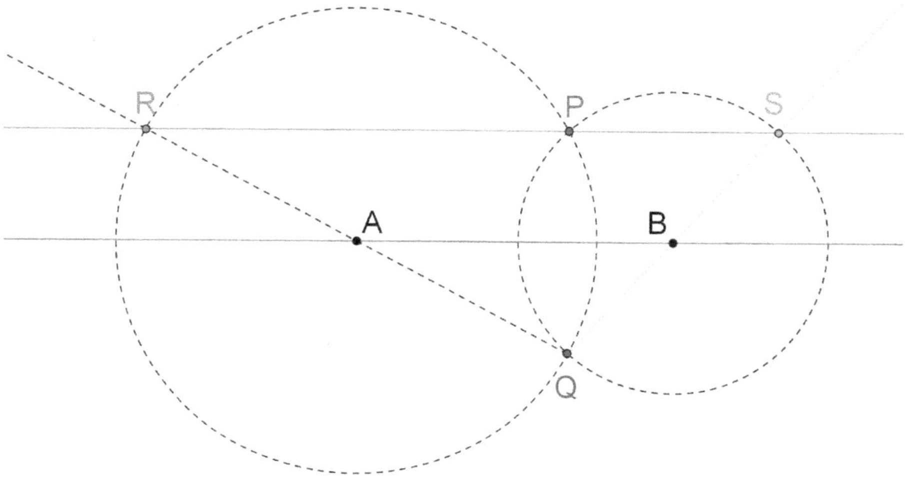

Figure I.2.6: The parallel using similar triangles.

Construction I.2.4 (The parallel obtained from similar triangles). *Choose any two points on the given line and draw circles with these centers through the given point P. They intersect in a second point Q, too, which is the the reflection image of P across the line l.*

The segment PQ intersects the given line in a point O, around which we draw a circle through points P and Q. It intersects the given line l in points A and B. We draw the rays from point Q through points A and B, and circular arcs through these points through point Q. We get the intersection points R and S. The line $RS \parallel l$ is the parallel to line l through point P. All three points P, R and S lie on the parallel, providing a further check.

Problem I.2.10. *Given is a line l and a point P not lying on the line. We want to construct the parallel to line l through the point P. Provide drawing with notations to illustrate the construction I.2.4 given above.*

Construction I.2.5 (The parallel using the harmonic quadrilateral). *We draw a circle with center any point B on the line l, and obtain the diameter AC on this line.*

We choose any point D on the line PC and draw lines DA and DB. Let point E

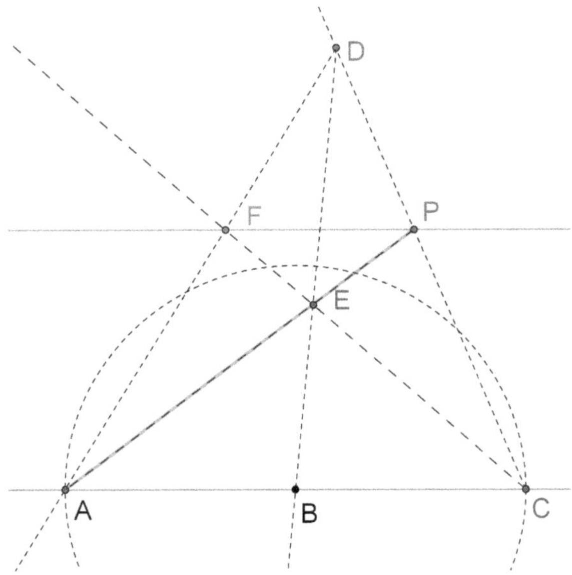

Figure I.2.7: The parallel using the harmonic quadrilateral.

be the intersection of lines AP and DB. Let point F be the intersection of lines EC and DA. The line PF is the parallel to the given line l through point P.

Problem I.2.11. *Given is a line l and a point P not lying on the line. We want to construct the parallel to line l through the point P. Provide drawing with notations to illustrate the construction I.2.5 given above.*

I.2.3 Dividing a Segment into any Number of Congruent Parts

Problem I.2.12. *The subdivision of a segment into three congruent parts, can be done as in the figure on page 54. Give a stepwise description of this construction.*

Construction I.2.6 (Subdivision using parallels). *Given is the segment AB, to be subdivided into n congruent parts. We draw any ray with vertex A and transfer onto it n congruent segments $AA_1, A_1A_2, \ldots, A_{n-1}A_n$. We draw the parallels to line A_nB through all intermediate points $A_1 \ldots A_{n-1}$. They intersect the given segment AB in the required division points $B_1 \ldots B_{n-1}$ such that segments*

$$AB_1, B_1B_2, \ldots, B_{n-1}B$$

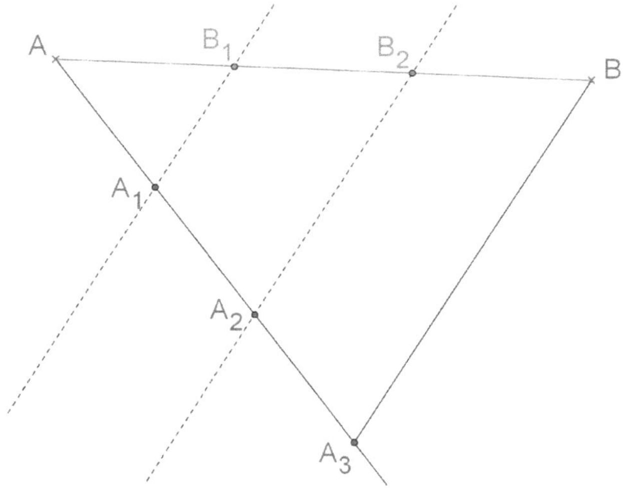

Figure I.2.8: Trisection using parallels.

are the n congruent parts of the given segment AB.

Remark. The required parallels are easily constructed by at first dropping the perpendicular p from A onto line A_nB, and then dropping perpendiculars onto line p from the intermediate points $A_1 \dots A_{n-1}$.

Take two equilateral triangles with the common side AB. Together they give the rhombus $\square ACBF$ as shown in the figure on page 55. We draw the long diagonal CF. Let M and N be the midpoints of segments AC and AF. The segment BM intersects the long diagonal CF in point D. The segment BN intersects the long diagonal CF in point E. We see and explain the following facts:

- At vertex B we get the angles $\angle CBM = 30°$, $\angle MBN = 60°$ and $\angle NBF = 30°$.
- The points D and E trisect the long diagonal CF.

Construction I.2.7 (Trisection using triangles). *Given is the segment AB, to be trisected. One constructs two congruent isosceles $\triangle ABC$ and $\triangle ABD$, lying on different sides of AB. Let M and N be the midpoints of segments AC and BC. Let P and Q be the intersections of segment AB with segments DM and DN, respectively.* Claim: *The segments AP, PQ and QB are congruent. Each one measures one third of AB.*

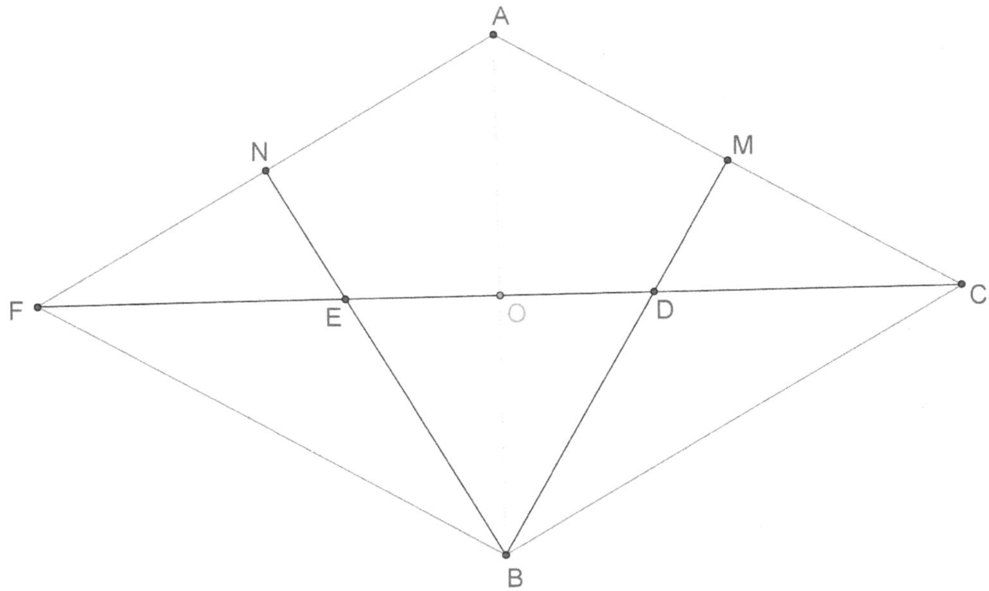

Figure I.2.9: How to trisect a segment but not an angle.

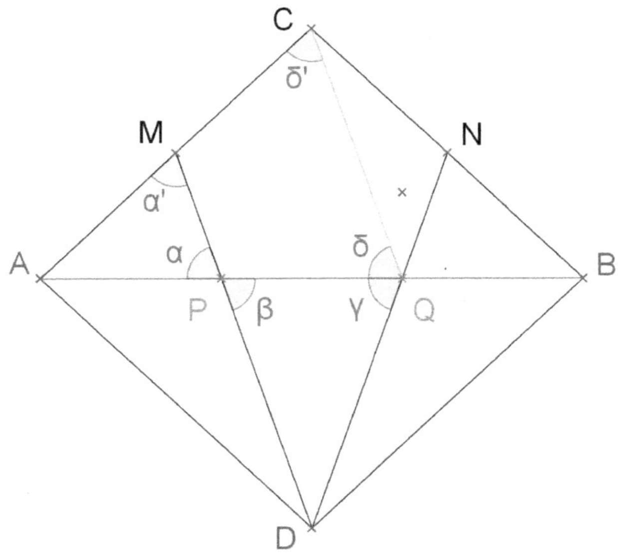

Figure I.2.10: Trisection of a segment

Problem I.2.13. *Complete the following proof of that claim. Indicate is yellow, which part of the argument still works, in green, which part does no longer work in hyperbolic geometry.*

Proof. We show that angles $\alpha = \angle APM, \beta = \angle QPB, \gamma = \angle PQD, \delta = \angle PQC$ are all congruent.

Indeed $\alpha \cong \beta$, because they are vertical angles.

Next $\triangle AMD \cong \triangle BND$, by SAS congruence with angles at vertices A and B and the adjacent segments matched.

Next $\triangle APD \cong \triangle BQD$ by ASA congruence with segments AD and BD and the adjacent angles matched. Hence $\angle APD \cong \angle BQD$, and $\beta = \gamma$, because supplements of congruent angles are congruent.

Next $\triangle APM \cong \triangle BQN$ by SAS congruence with angles at vertices A and B and the adjacent segments matched. Hence especially $AP \cong BQ$.

Next $\triangle AQC \cong \triangle AQD$, by SAS congruence, with angle at vertex A and adjacent segments matched. Hence $\gamma = \delta$.

Those congruences allow to get the claims $\alpha = \delta$ and $AP \cong BQ$. Hence by Euclid's Proposition I.27, the lines MD and CQ are parallel. The reasoning up to this point is valid in neutral geometry. We can carry the argument further in Euclidean geometry.

Indeed Euclid I.29 yields $\angle AMP \cong \angle ACQ$. Hence $\triangle AMP$ and $\triangle ACQ$ have all angles pairwise congruent, and are similar. By Euclid VI.4, the sides of equiangular triangles are proportional. Hence we get the proportion

$$\frac{|AC|}{|AM|} = \frac{|AQ|}{|AP|} \tag{I.2.1}$$

Since $MC \cong AM$ and $AC = AM + MC$ by construction, the ratios in (I.2.1) are both equal to 2. Hence $|AQ| = 2|AP|$, too. Now $AQ = AP + PQ$ implies that the segments AP and PQ are congruent. Thus all three segments AP, PQ and QB have been shown to be congruent, and hence all three have $\frac{1}{3}$ of the length of AB. ☐

Problem I.2.14 (Using only equilateral triangles). *Draw only two circles of radius AB, and then complete the construction just with a straightedge.*

Construction I.2.8. *Begin by constructing two congruent equilateral $\triangle ABC$ and $\triangle ABD$, lying on different sides of AB.*

Line DA intersects the left circle in a second point E. Line DB intersects the right circle in a second point F. Drawn segment BE, which intersects AC in the midpoint

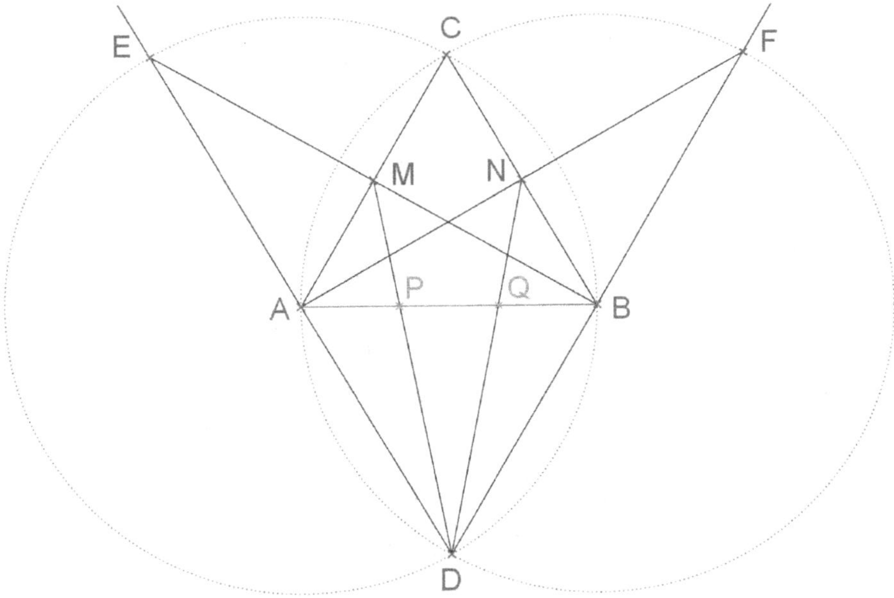

Figure I.2.11: Completed trisection with equilateral triangles

M. Draw segment AF, which intersects BC in the midpoint N. Draw segments MD and ND. They intersect the given segment AB in points P and Q.
Claim*: The segments AP, PQ and QB are congruent. Each one measures one third of AB.*

Problem I.2.15. *For clever students: In hyperbolic geometry, does the middle segment get shorter, or longer than the other two segments?*

Answer. In hyperbolic geometry, the middle segment gets shorter than the two other segments. Here is a reasoning for the special case that $\triangle AQC$ is isosceles. In that case $\delta \cong \delta' := \angle ACQ$, because they are base angles of an isosceles triangle.

Next, I use the defect of the angle sum $2R - \alpha - \beta - \gamma$, which is proportional to the area of a triangle. $\triangle APM$ is part of $\triangle AQC$, and hence has smaller area, and smaller defect.

Let $\alpha' := \angle AMP$. The defect of $\triangle AMP$ is less than the defect of the larger $\triangle ACQ$. Comparison of the defects leads to $\alpha + \alpha' > \delta + \delta'$ and $\alpha' > \delta'$. Euclid I.19 says:

```
"If one angle of a triangle is greater than another, then the
  side opposite to the angle is greater than the other."
```

Now, we have found that $\angle AMP = \alpha' > \delta' = \delta = \alpha = \angle APM$, and we can use Euclid I.19 in $\triangle APM$. Hence $AP > AM$. Now $AQ \cong AC$ implies $PQ = AQ - AP < AC - AM = MC \cong AM < AP$.

Remark. In hyperbolic geometry, the problem to trisect a given segment cannot be solved with straightedge and compass. Indeed, the problem leads to the Delian problem to construct $\sqrt[3]{2}$, which is impossible to do with straightedge and compass.

I.2.4 Some Triangle Constructions

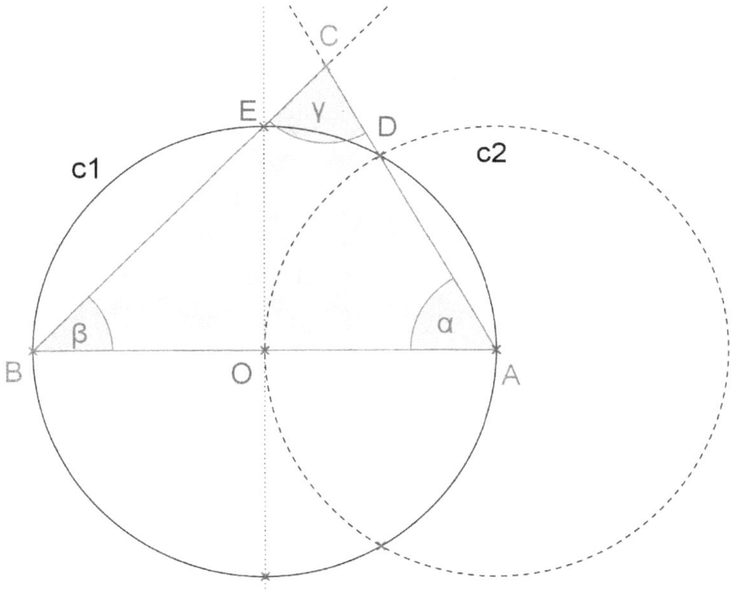

Figure I.2.12: What are the angles of $\triangle ABC$?

Problem I.2.16 (Another special triangle). *Describe the construction done in the figure on page 58. Explain how you determine the three angles of the triangle* $\triangle ABC$.

Answer. On an arbitrary segment OA, an equilateral triangle $\triangle OAD$ is erected. This can be done as described in Euclid I.1. We find the intersection point D of a circle around A through O with a circle around O through A.

Let B be the second endpoint of diameter AOB. At point O, we erect the perpendicular onto this diameter. Let E be the intersection of the perpendicular with

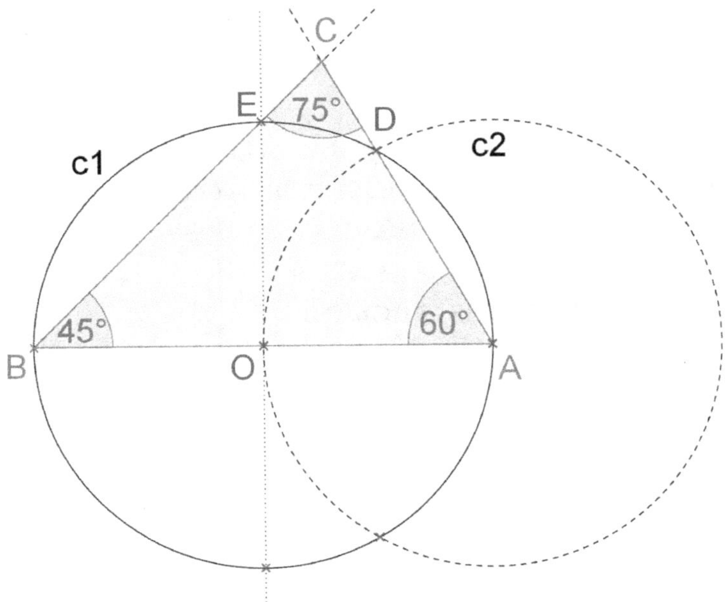

Figure I.2.13: Construction of a triangle with angles of $45°, 60°$ and $75°$.

the right circle, lying on the same side of AB as point D. The rays $\overrightarrow{AD}$ and $\overrightarrow{BE}$ intersect in point C.

We thus get a triangle $\triangle ABC$ with the angles $60°, 45°$ and $75°$ at its vertices A, B and C. The angle $\angle CAB = 60°$ is obtained from the equilateral triangle $\triangle OAD$. The angle $\angle ABC = 45°$ is obtained from the right isosceles triangle $\triangle OBE$. Finally, one calculates the third angle $\angle BCA = 180° - 60° - 45°$ by the angle sum of triangle $\triangle ABC$.

Problem I.2.17 (A construction using an altitude). *Using Euclid III.21, construct a $\triangle ABC$ with the three following pieces given: side $c = AB = 6$, opposite angle $\gamma = \angle BCA = 30°$, and altitude $h_c = 5$.*
(h_c is the altitude dropped from vertex C onto the opposite side AB).

Question (**a**). Do the construction and measure your angles α and β.

Answer.

Question (**b**). Describe the steps for your construction.

Construction I.2.9. *Draw side $AB = 6$ and its perpendicular bisector p. Let M be the midpoint of AB. The center O of the circum circle lies on the perpendicular*

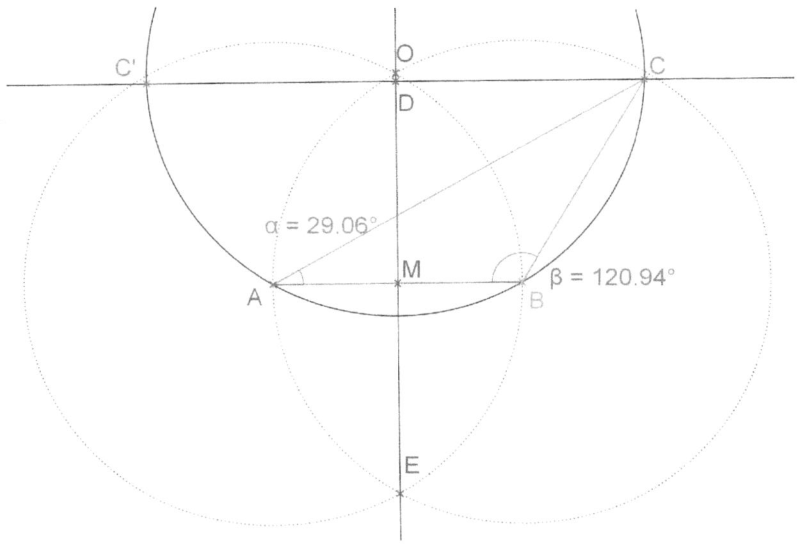

Figure I.2.14: A triangle construction

bisector. The center angle is double the circumference angle γ. Hence $\angle AOB = 2\gamma = 60°$, and $\angle AOM = 30°$. For the example given, the point O is especially easy to find because the $\triangle AOB$ is equilateral. Next, we draw the circle around O through A and B. This is the circum circle of $\triangle ABC$, on which vertex C lies.

Secondly, vertex C lies on a parallel q to AB of distance $|MD| = h_c = 5$, because of the given altitude. Hence vertex C is an intersection point of this parallel with the circum circle. One may choose any one of these two intersection points.

Remark. One can check that, in Euclidean geometry, the triangle $\triangle ABC$ is <u>not</u> isosceles: Indeed, the isosceles triangle $\triangle ABC'$ with $AB = BC' = 6$ and $\gamma = 30°$ has the altitude

$$h'_a = 6 \cdot \frac{\sqrt{3}}{2} = \sqrt{27} > 5$$

Hence point D lies between M and O, and $\alpha < 30°$.

Problem I.2.18 (A further construction using an altitude). *Using Euclid III.21 as well as the tangent to a circle, construct a triangle $\triangle ABC$ with the three following pieces given: side $c = AB = 6$, opposite angle $\gamma = \angle BCA = 30°$, and altitude $h_a = 4$. (h_a is the altitude dropped from vertex A onto the opposite side BC).*

Question **(a)**. Do the construction and measure your angles α and β.

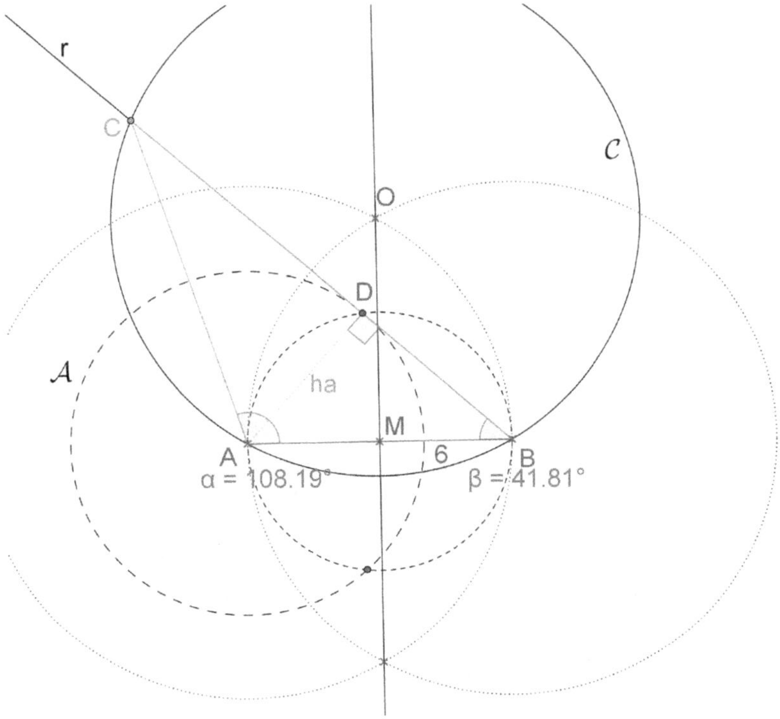

Figure I.2.15: A triangle construction

Answer. The construction is shown in the figure on page 61. The angles are measured to be $\alpha = 108.19°$ and $\beta = 41.81°$.

Question (**b**). Describe the steps for your construction.

Construction I.2.10. *Draw side $AB = 6$ and its perpendicular bisector p. Let M be the midpoint of AB. The center O of the circum circle lies on the perpendicular bisector. The center angle is double the circumference angle γ. Hence $\angle AOB = 2\gamma = 60°$, and $\angle AOM = 30°$. For the example given, the point O is especially easy to find because the $\triangle AOB$ is equilateral. Next, we draw the circle around O through A and B. This is the circum circle $\mathcal{C}$ of $\triangle ABC$, on which vertex C lies.*

Next we draw the circle $\mathcal{A}$ of radius $h_a = 4$ around point A. The triangle side BC lies on the tangent r to this circle from point B. In the figure on page 61, the tangent r has been constructed using the Thales circle with diameter AB. In the figure on page 62, the tangent r has been constructed by Euclid's method.

Vertex C is an intersection point of this tangent r with the circum circle $\mathcal{C}$. In the upper half plane of AB one gets only one solution.

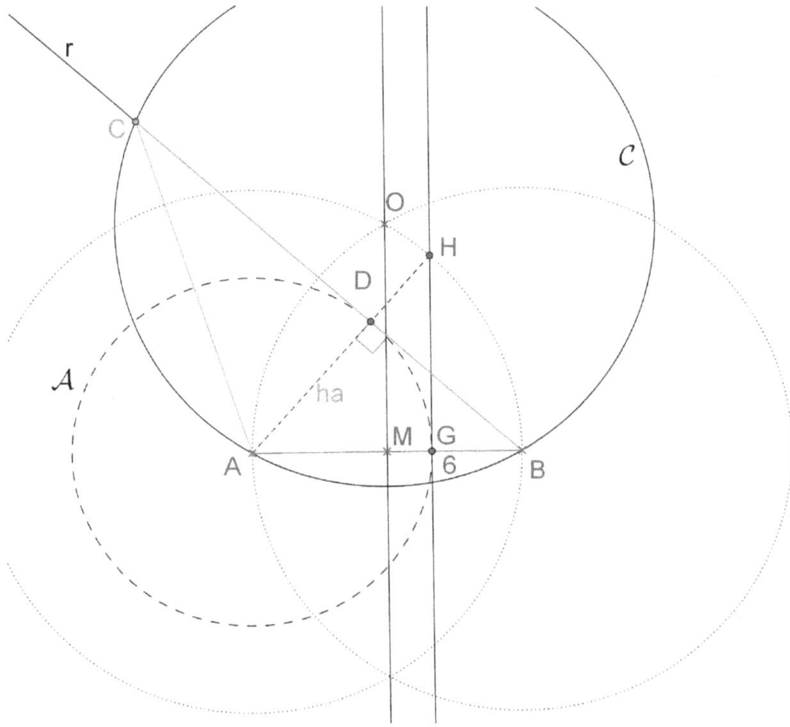

Figure I.2.16: A triangle construction

Problem I.2.19. *Given is a triangle $\triangle ABC$ with its three altitudes. Find in this figure three pairs of congruent angles at the vertices A, B, C.*

Hint: *Draw the circle with a triangle side AB as diameter. Because of the converse Thales' theorem, two foot-points of altitudes lie on this circle. Now use Euclid III.21 to find congruent circumference angles.*

Answer. We draw a semicircle with a triangle side AB as diameter. Because of the converse Thales' theorem, two foot-points D and E of altitudes lie on this circle. By Euclid III.21 , the arc between these two foot points has congruent circumference angles at the vertices A and B. From the triangle sides BC and CA, we get two further pairs of congruent angles.

Problem I.2.20 (A construction using an altitude). *Using Euclid III.21, construct a triangle $\triangle ABC$ with the three following pieces given: side $c = AB = 6$, opposite angle $\gamma = \angle BCA = 60°$, and altitude $h_c = 4$.*
(h_c is the altitude dropped from vertex C onto the opposite side AB).

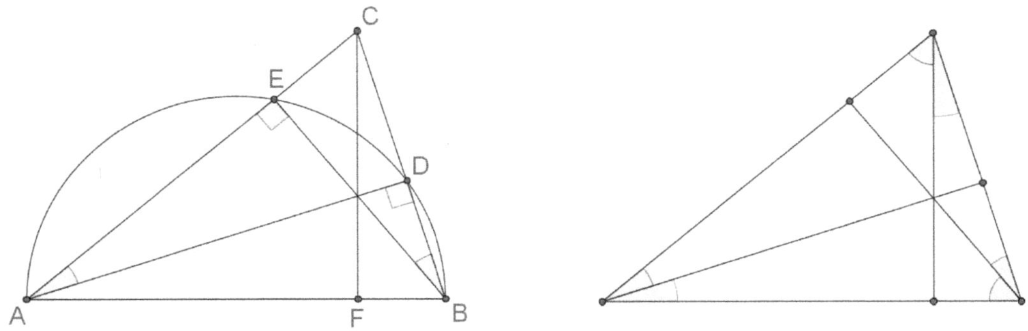

Figure I.2.17: Congruent angles produced by the altitudes of a triangle.

Question (**a**). Do the construction and measure your angles α and β.

Answer. The construction is shown in the figure on page 64. The angles are measured to be $\alpha = 35.49°$ and $\beta = 84.51°$.

Question (**b**). Describe the steps for your construction.

Answer. Draw side $AB = 6$ and its perpendicular bisector p. Let M be the midpoint of AB. The center O of the circum circle lies on the perpendicular bisector p. Because the value $\gamma = 60°$ is given in the example, we construct at first the equilateral triangle $\triangle AEB$. Next we need the circum circle $\mathcal{C}$ of the equilateral triangle $\triangle ABE$. As follows from Euclid III.21, the vertex C of the triangle $\triangle ABC$ to be constructed lies on the circle $\mathcal{C}$, too .

To get the center of circle $\mathcal{C}$ we draw the circle around E through A and B. It intersects the circle around B through A and E at the points A and R. The line AR is the perpendicular bisector of segment BE, and intersects the perpendicular bisector p of segment AB at point O, which is the center of the circum circle $\mathcal{C}$ of the equilateral triangle $\triangle ABE$. The circum circle $\mathcal{C}$ is now easily drawn.

Secondly, vertex C lies on a parallel q to AB of distance $|MD| = 4$, because of the altitude is given to be $h_c = 4$. The vertex C is an intersection point of this parallel with the circum circle $\mathcal{C}$. In the upper half planenn of AB one gets two solutions. One may choose any one of these two intersection points.

I.2.5 Quadrilaterals

Problem I.2.21. *In Euclidean geometry, find convex quadrilaterals with the interior angles of* $30°, 60°, 120°, 150°$.

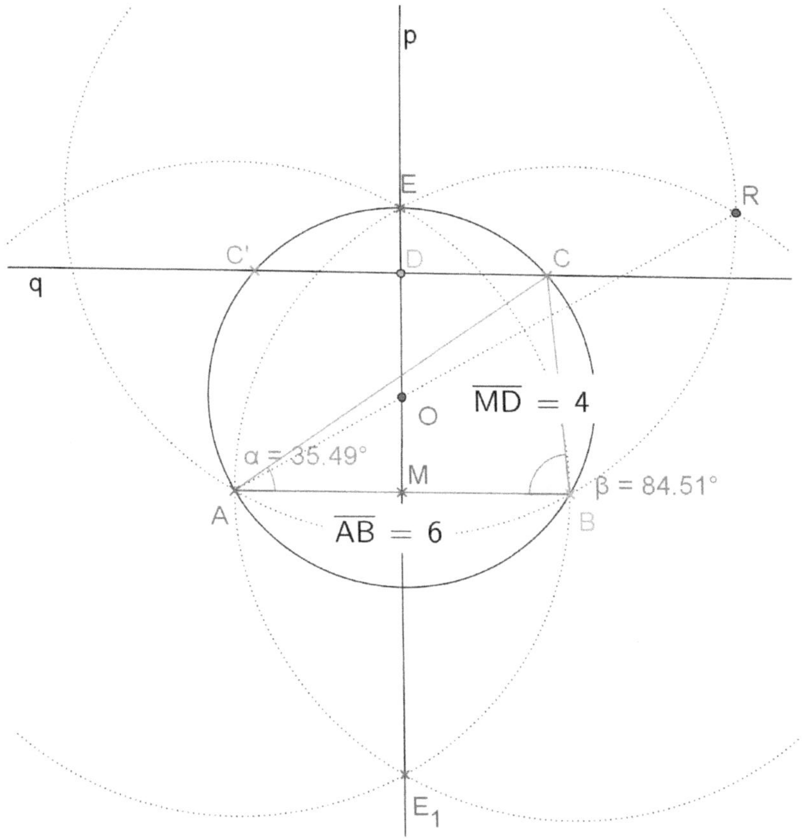

Figure I.2.18: A triangle construction

(i) *Convince yourself that such quadrilaterals all are either trapezoids or circular quadrilaterals.*

(ii) *We assume additionally that two adjacent sides are congruent. I do not count separately the reflexion symmetric instances, nor congruent or similar instances,— how many different shapes can these quadrilaterals have? Convince yourself that there exist five different shapes for trapezoids, and one shape for a circular quadrilateral.*

(iii) *Give accurate drawings, using a protractor and straightedge, or geogebra.*

Answer. **(i)** If the angles 30° and 150° are adjacent, one gets a trapezoid, if these two angle are opposite angles, one gets a circular quadrilateral.

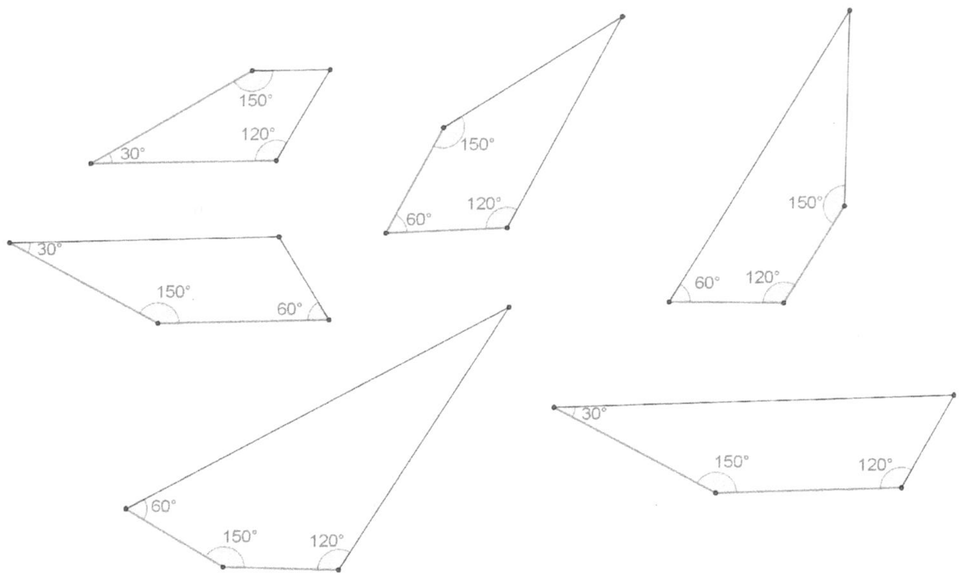

Figure I.2.19: Six quadrilaterals.

(ii) We assume additionally that two adjacent sides are congruent.

- In case the angle of 30° includes two congruent sides, the adjacent angle cannot be 60°. Hence the adjacent angles are 120° and 150°.
- In case the angle of 60° includes two congruent sides, the adjacent angle cannot be 30°. Hence the adjacent angles are 120° and 150°.
- In case the angle of 120° includes two congruent sides, the adjacent angle cannot be 30°. Hence the adjacent angles are 60° and 150°.
- In case the angle of 150° includes two congruent sides, all three choices of the adjacent angles among 30°, 60° or 150° are possible.

Problem I.2.22. *Given are three points A, B and C. On which curve $\mathcal{D}$ lies the forth vertex of a quadrilateral $\square ABCD$, for which*

(a) *the opposite angles have sum $\alpha + \gamma = 120°$,*

(b) *the opposite angles have sum $\alpha + \gamma = 270°$.*

Again, the vertices A, B, C, D are named in alphabetic order as they follow each other around the quadrilateral.

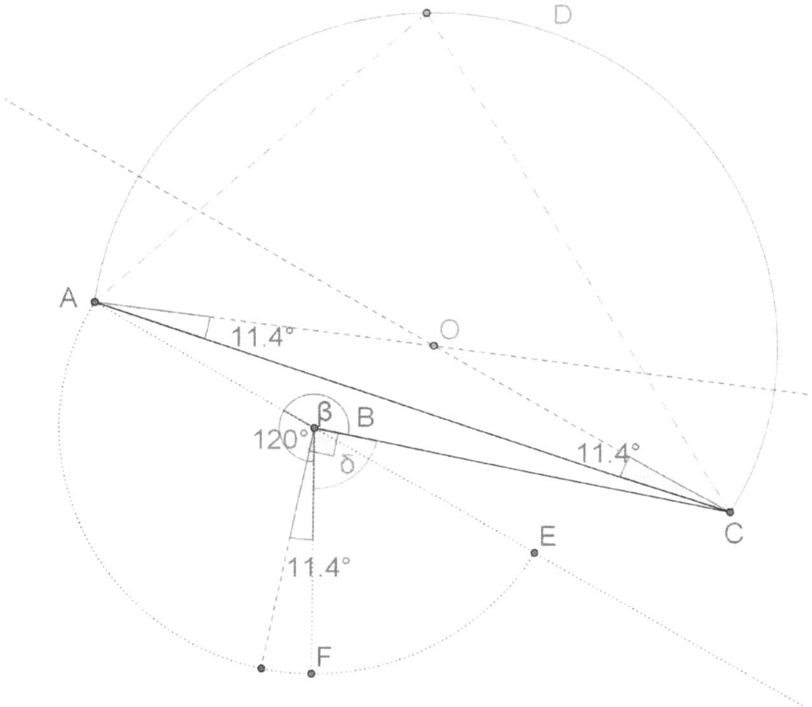

Figure I.2.20: A quadrilateral with $\alpha + \gamma = 120°$ and given vertices A, B, C.

Actually, the curve $\mathcal{D}$ is a circle. In two instances for the examples (a) *and* (b) *above, construct this circle, and especially its center.*

Answer. Because the angle sum of a quadrilateral is 360°, the angle at the forth vertex is $\delta = 360° - \alpha - \beta - \gamma$. In case (a), we get $\delta = 360° - \beta - 120°$. This angle can easily be constructed at vertex B since $\beta = \angle ABC$ is given. In the figure on page 66, the equilateral triangle $\triangle BEF$ is used to obtain the angle $\angle ABF = 120°$ and hence the angle $\angle FBC = \delta$. The complementary angle $90° - \delta$ is transferred to the rays $\overrightarrow{AC}$ and $\overrightarrow{CA}$. The intersection of these two rays is the center O of the circle through A and C on which the forth vertex D lies. Indeed, the center angle is $\angle AOC = 2\delta$ and, by Euclid III.21, the circumference angle is $\angle ADC = \delta$ as required.

In case (b), the angle at the forth vertex is $\delta = 360° - \alpha - \beta - \gamma = 90° - \beta$. Hence we transfer angle β to the rays $\overrightarrow{AC}$ and $\overrightarrow{CA}$. The intersection of these two rays is the center O of circle to be constructed.

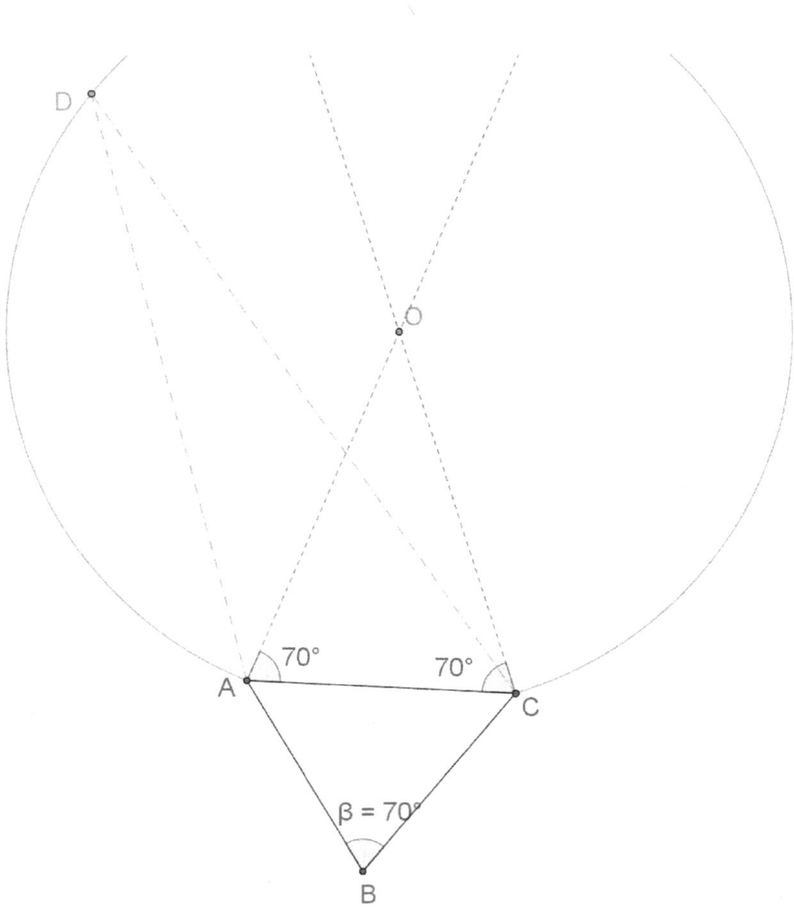

Figure I.2.21: A quadrilateral with $\alpha + \gamma = 270°$ and given vertices A, B, C.

Part II

Toward a Modern View

II.3 Hilbert's Axiomatic Approach, a Short Review

II.3.1 Logic

This subsection is too short to do justice to its topic. A comprehensive introduction into mathematical logic may be found in the book *"A Tour through Mathematical Logic"* by Robert S. Wolf, see [20].

By a *statement* is meant any sentence for which it is meaningful to consider its truth or falsehood. [3] Thus any statement can be either true or false,— but a statement is not automatically assumed to be true. Often, I shall put quotation marks around statements in order to remind the reader to this important remark. Let A and B denote any statements. The *negation* of a statement is true if and only if the statement itself is false. The negation of formula A is denoted by ¬A or not(A) or simply not A.

By $A \equiv B$, we denote a <u>true</u> equivalence. Thus A ≡ B means that it is true that either formulas A and B are both true, or formulas A and B are both false. [4]

A *formal implication* A ⇒ B is defined by a truth table—without assuming or presupposing any causal relation between the statements A and B. According to such a definition, the implication A ⇒ B becomes false if and only if statement A is true and statement B is false. Thus

$$\neg(\mathrm{A} \Rightarrow \mathrm{B}) \equiv \mathrm{A\,and}\,\neg\mathrm{B}$$
$$\mathrm{A} \Rightarrow \mathrm{B} \equiv \neg\mathrm{A\,or\,B}$$

The statements A ⇒ B and A ⇒ ¬B can both be true simultaneously. They are not negations of each other. Indeed

$$(\mathrm{A} \Rightarrow \mathrm{B}) \text{ and } (\mathrm{A} \Rightarrow \neg\mathrm{B}) \equiv \mathrm{A} \text{ is false}$$

The *converse* of the statement A ⇒ B is the statement B ⇒ A. Remember: The converse of a true statement may be either true or false.

The *contrapositive* of the statement A ⇒ B is the statement ¬B ⇒ ¬A.

Lemma II.3.1. *Any statement is equivalent to its contrapositive. As a formula:*

$$\mathrm{A} \Rightarrow \mathrm{B} \equiv \neg\mathrm{B} \Rightarrow \neg\mathrm{A}$$

---

[3] This is a rather ad hoc definition, but I cannot find any better one.

[4] It would be good to have a different notation for an arbitrary equivalence, which may be true or not. I do not suggest to give the symbol ⇔ such a meaning. Above all, the reader should remember that it is unusual and dangerous to write down false statements!

> geometry and the investigation of their mutual connections has been the subject of many excellent treatises in the mathematical literature. The problem in question is basically a logical analysis of our spatial imagination.
>
> The present—meaning Hilbert's—investigation is a new attempt to set up a *complete* and *as simple as possible* system of axioms. And, furthermore, to deduct from them the most important geometric theorems, such that the meaning and importance of the different axioms and their consequences become clear.
>
> Hilbert's introduction to *"Foundations of Geometry"* (1899)

Hilbert's axioms

As suggested in this paragraph, Hilbert introduces the modern habit to develop the consequences of the different groups of axioms immediately after introducing them. But—because I guess it is more convenient for the reader—we shall now state the complete axiomatic system at once, at the beginning. I have included all axioms, even those only needed for three dimensional geometry.

0. Undefined elements and relations

Elements:

- A class of undefined objects called *points*, denoted by $A, B, C, \ldots$.
- A class of undefined objects called *lines*, denoted by $a, b, c, \ldots$.
- A class of undefined objects called *planes*, denoted by $\alpha, \beta, \gamma, \ldots$.

Relations:

- *Incidence* (being *incident, lying on, containing*)
- *Order* (lying *between*) (for points on a line)
- *Congruence* (for segments and angles)

Remark. Planes are only needed to include three dimensional geometry.

I. Axioms of incidence

I.1 For two points A and B there exists a line that contains each of the points A, B.

I.2 For two [different] points A and B there exists no more than one line that contains each of the points A, B.

I.3a On any line, there exist at least two points.

I.3b There exist at least three points that do not lie on a line.

Remark. We did separate axiom I.3 into I.3a and I.3b, in order to stress there is no direct logical connection between the two sentences intended.

Remark. Additional axioms I.4 through I.8 are only needed for three dimensional geometry.

I.4 For any three points A, B, C [that do not lie on the same line], there exists a plane α that contains each of the points A, B, C.

I.4a For every plane there exists a point which it contains.

I.5 For any three points A, B, C that do not lie on the same line, there exists no more than one plane that contains each of the points A, B, C.

I.6 If two different points $A \neq B$ of a line a lie on a plane α, then every point of line a lies in the same plane α.

Remark. In this case, we say that the line a lies in the plane α.

I.7 If two planes α, β have a point A in common, then they have at least one more point B in common.

I.8 There exist at least four points which do not lie in a plane.

II. Axioms of order

II.1 If a point B lies between a point A and a point C, then the points A, B, C are three distinct points of a line, and B lies between C and A.

II.2 For two points A and C, there also exists at least one point D on the line AC such that C lies between A and D.

II.3 Of any three points on a line there exists no more than one that lies between the other two.

DEFINITIONS: Segment, point of segment, interior and exterior of segment, ray, half plane, triangle. [5]

Remark. A segment AB is assumed to have two different endpoints A and B.

II.4 (Pasch' Axiom) Let A, B, C be three points that do not lie on a line and let a be a line in the plane ABC which does not meet any of the points A, B, C. If the line a passes through a point of the segment AB, it also passes through a point of the segment AC, or through a point of the segment BC.

III. Axioms of congruence

III.1 If A, B are two points on a line a, and A' is a point on the same or another line a', then it is always possible to find a point B' on a given side of the line a' through A' such that the segment AB is congruent to the segment $A'B'$. In symbols $AB \cong A'B'$.

III.1b For any two different points A and B, the segments AB and BA are congruent. [6]

III.2 If a segment $A'B'$ and a segment $A''B''$ are congruent to the same segment AB, then segment $A'B'$ is also congruent to segment $A''B''$.

III.3 On a line, let AB and BC be two segments which except for B have no point in common. Furthermore, on the same or another line a', let $A'B'$ and $B'C'$ be two segments which except for B' also have no point in common. In that case,

$$\text{if} \quad AB \cong A'B' \quad \text{and} \quad BC \cong B'C' \,, \quad \text{then} \quad AC \cong A'C'$$

DEFINITION: Angle.

III.4a Let $\angle(h, k)$ be an angle in a plane α and h' a ray in a plane α' that emanates from the point O'. Then there exists in the plane α' one (and only one) ray k' such that the angle $\angle(h, k)$ is congruent to the angle $\angle(h', k')$ and at the same

---

[5] lateron: quadrilateral, polygon, side of polygon, vertex of polygon, closed and open polygon, simple polygon.

[6] I have added the item (III.1b) to Hilbert's axioms. This is necessary in my opinion.

time all interior points of the angle $\angle(h', k')$ lie on the given side of h'. This means that

$$\angle(h, k) \cong \angle(h', k')$$

III.4b The ray k' in (III.4a) is unique.

III.4c Every angle is congruent to itself

$$\angle(h, k) \cong \angle(h, k)$$

III.4d Every angle is congruent to the angle with the legs switched

$$\angle(h, k) \cong \angle(k, h)$$

III.5 If for two triangles $\triangle ABC$ and $\triangle A'B'C'$ the congruences

$$AB \cong A'B' \,, \; AC \cong A'C' \,, \;\; \angle BAC \cong \angle B'A'C'$$

hold, then the congruence $\angle ABC \cong \angle A'B'C'$ is also satisfied.

IV. Axiom of parallelism

IV.1 Let a be any line and A a point not on a. Then there exists at most one line in the plane determined by line a and point A that passes through A and does not intersect a.

V. Axioms of continuity

V.1 (Axiom of Archimedes) If AB and CD are any segments, then there exists a number n such that n segments congruent to CD constructed contiguously from A, along a ray from A through B, will pass beyond B.

V.2 (Axiom of completeness) An extension of a set of points on a line, with its order and congruence relations existing among the original elements as well as the fundamental properties of line order and congruence that follow from Axioms I-III and from (V.1), is impossible.

Definition II.3.1 (Hilbert plane). A *Hilbert plane* is any model for two-dimensional geometry where Hilbert's axioms of incidence (I.1)(I.2)(I.3a)(I.3b), order (II.1) through (II.4), and congruence (III.1) - (III.5) hold.

Neither the axioms of continuity—Archimedean axiom and the axiom of completeness, nor the parallel axiom need to hold for an arbitrary Hilbert plane.

Definition II.3.2 (Pythagorean plane). A *Pythagorean plane* is a Hilbert plane for which the axiom of parallelism (IV.1) holds.

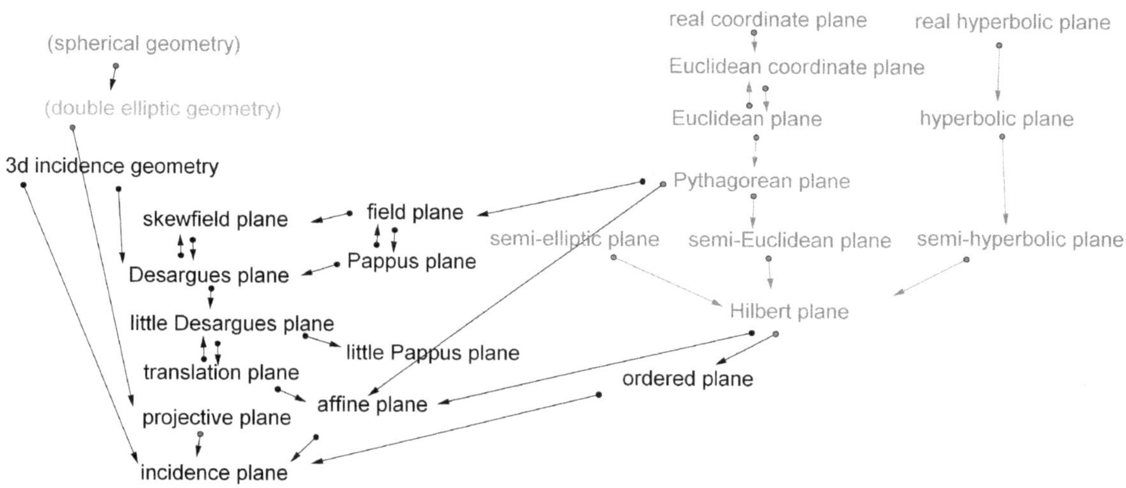

Figure II.3.1: Logical relations of geometries

II.3.3 Frege's Critique and Hilbert's Answer

The book *Foundations of Geometry* was discussed and critized by Frege in an exchange of letters between the two scholars. Between October 1895 and November 1903, we know of nine letters having been exchanged. In the letter of December 1899 Hilbert writes:

> I still have to talk about one objection. You tell me, my notions, for example "point", "between" have not been fixed uniquely—for example the notion "between" is grasped (gefasst) differently on p.20 and there a point is a pair of numbers—.

> Yes, it is indeed self evident that every theory is only a framework or scheme of notions together with the necessary relations between them, and the basic elements (Grundelemente) can be thought of in any arbitrary manner. When I imagine as my points [their relation of order, my lines] any arbitrary system of things, for example the system: love, law, chimney sweeper...,
>
> I imagine [this system] and then I think of all my axioms as connections between these things; henceforth my theorems, for example the theorem of Pythagoras, are valid about these things, too.
>
> In other words: every theory can always be applied to infinitely many systems of basic elements (Grundelemente). One has indeed only to apply a one-to-one invertible transformation and agree [to postulate] the same corresponding axioms for the transformed basic elements. In fact this state of affairs is used frequently, for example in projective geometry as principle of duality, and by me in proofs of independence.

On another occasion, Hilbert made the point that his proofs should stay completely correct even if the words "point", "line", and "plane" were replaced throughout by "table", "chair", and "beer mug." In the two last letters of September 1900 and November 1903, Hilbert formulates his point of view concisely. In September 1900 Hilbert answers to Frege:

> It is really my opinion that a notion can only be logically clarified by its relations with other notions. These relations, formulated in distinct assertions, I call axioms. Hence I arrive at considering the axioms (together with the names for the notions) as the definitions of the notions. These convictions did not arise for me out of pure fancy, but I see myself pushed to them as a necessity for the rigorous and logical building of a theory. I have arrived at the conviction that subtle facts in mathematics and the natural sciences can be treated with certainty only in this way—otherwise one only turns oneself in cycles.

II.3.4 Drawbacks and Lacuna of Hilbert's Foundations

I want to ask, which topics are still not included, and what are the results achieved elsewhere about these topics. Of course I look at first at the present manuscript, after that beyond at the worldwide research.

The first subject I miss in Hilbert's foundation, are the axioms and theorems dealing with circles. Researchers like Greenberg and Hartshorne have meanwhile introduced the *line-circle intersection property* V.12.7 and the *Circle-circle intersection property* V.12.8 as the relevant axioms about the intersection of circles with lines, and of circles with circles. Several interesting theorems can be investigated already in neutral geometry. There is the remarkable three-circle theorem about of the threefold intersection of their three common chords, which holds even in neutral geometry. By the way, the figure is depicted on the cover of Robin Hartshorne's book [7] *Geometry: Euclid and Beyond.* Here, it would be nice to have a proof totally in neutral geometry. Presently, I know only a proof done at first in Euclidean geometry, and in a second step extended to hyperbolic geometry by means of the Poincaré disk model.

The *Uniformity Theorem* 6 allows to classify all Hilbert planes into one of three types, depending solely on the angle sum of triangles. This is a sharpening of Legendre's Second Theorem. Hilbert includes the two Legendre Theorems, but only mentions the more general Uniformity Theorem, giving credit to his student Max Dehn. Actually, the result may have been known earlier, A complete proof is contained in Robin Hartshorne's book *Geometry: Euclid and Beyond* [7] and in the present manuscript.

Greenberg [6] has recently stressed the equivalence of the Euclidean parallel axiom with the angle sum of the triangle being two right angles together with *Aristole's Angle Unboundedness Axiom* V.12.11. Proclus' Theorem 7 clarifies and sharpens very old ideas of Proclus about the parallel postulate.

A corresponding result in hyperbolic geometry is much deeper and harder to prove. Here it would be a progress to have a proof totally in neutral geometry. The current proof mentioned by Greenberg [6] in his article *Old and new results in the foundations of elementary plane Euclidean and Non-Euclidean geometries*, is based on rather sophisticated elaborations of the Klein disk model of hyperbolic geometry.

II.3.5 General Remark About Models in Mathematics

The meaning and usage of the notion "model" in the natural sciences on the one side, and mathematics on the other side, are different. For example, the astronomer says that Newton's theory of gravitation is a model for the solar system. The mathematician says that the solar system is a model of Newton's theory of gravitation. [7]

In the natural sciences, one has to begin with a complicated reality. One wants to

[7] Another consequence of these matters: A scientific model can only be falsified. A mathematical model can be verified

extract some salient features, and describe or explain them in hopefully more simple mathematical terms. The scientist calls such a mathematical description a model.

On the other hand, the mathematician already begins with a theory, which often can be pinpointed by the appropriate axioms. In a second step, one looks for other examples, problems or contexts to which the theory fits. To achieve such a fit, the abstract objects and relations of the theory have to be interpreted as specific notions occurring in the example at hand. Being interpreted in the context of the example, one has to ask and check whether the objects and relations from the abstract theory actually satisfy the axioms originally postulated. That can turn out to be true or not, depending on the properties of the example chosen. Furthermore, it can turn out that the same theory fits for quite different contexts.

In building a model, the mathematician has to think on two levels. The basic level is given by the specific example at hand and its accepted properties. In favorite circumstances, one needs only rudiments of set theory and the number system as accepted properties. One can think of this level as the "background ontology". The secondary level brings in the originally undefined objects and relations of an abstract, and often more modern, theory, which now get interpreted in terms of the basic level. If the axioms can be proved to hold for this specific example, we say that we have constructed a model for the abstract theory. In this way, one obtains a proof of relative consistency of the abstract theory.

II.4 Incidence Geometry

II.4.1 Elementary Propositions of Incidence Planes

Definition II.4.1 (Incidence plane). Any model for which Hilbert's axioms of incidence (I.1)(I.2)(I.3a)(I.3b) are valid is called an *incidence plane.*

Lemma II.4.1. *Any two different lines of an incidence plane have either no or exactly one point in common.*

Problem II.4.1. *Give a short explanation to convince yourself that Lemma II.4.1 does hold.*

Answer. Take any two different lines $l \neq m$. Assume towards a contradiction they intersect at two different points $A \neq B$.

By axiom (I.2), there exists at most one line through these two points. Hence $l = m$, contradicting the assumption that l and m are two different lines. Hence two different lines intersect at most in one point.

Definition II.4.2 (Parallel lines). Two different lines which lie in one plane and do not intersect are called *parallel.*

Question. Give at least two further useful formulations of Lemma II.4.1.

Answer. Here are three possible answers:

- Any two different lines lying in a plane and not parallel, have a unique point of intersection.
- If two lines have two or more points in common, they are equal.
- Two different lines have at most one point in common.

Lemma II.4.2. *In any incidence plane, there exist two different lines through every point.*

Proof. By axiom (I.3b), there exist at least three points that do not lie on a line. We call them A, B and C. Let any point P be given. The point P may be among the three points A, B, C or not. Thus the proof involves four points.

- In the case that point P is one of the three points A, B, C, we draw the three lines AB, BC and CA. These are three different lines, and two of them go through the given point P.

- In the case that point P is different from all three points A, B, C, we draw the three lines PA, PB and PC. At least two of them are different since A, B, C do not lie on a line.

In both cases we have obtained two different lines through the arbitrary point P. □

For two dimensional geometry, all lines and points are assumed to lie in the same plane. Referring to the parallel postulates, we need not state that the parallel to a given line through a given point actually lies in this plane. This leads to the following a bid simpler formulation:

Definition II.4.3 (Euclidean Parallel Postulate). For every line l and for every point P lying not on l, there *exists* a *unique* parallel m to l through point P.

But Hilbert's parallel axiom (IV) is a weaker assertion, it postulates only the uniqueness of the parallel to a given line through a given point.

Definition II.4.4 (Hilbert's Parallel Postulate). For every line l and for every point P lying not on l, there exists *at most* one parallel m to l through point P.

Proposition V.12.14 below implies existence of a parallel once the axioms of order and congruence are assumed. Hence Euclid's and Hilbert's parallel postulate turn out to be equivalent in Hilbert planes. But in the case of incidence planes, such an equivalence does not hold.

Definition II.4.5. We say that a two-dimensional incidence geometry for which the Euclidean Parallel Postulate holds, has the *Euclidean parallel property.*

Definition II.4.6. We say that a two-dimensional incidence geometry has the *elliptic parallel property* iff any two lines do intersect—necessarily at a unique point.

Definition II.4.7. We say that a two-dimensional incidence geometry has the *hyperbolic parallel property* iff for every line l and point P not on line l, there exist *at least* two parallels to line l through point P.

Lemma II.4.3 (Proclus' Lemma). *In any incidence plane where Hilbert's Parallel Postulate for plane geometry II.4.4 holds, the following statements are valid:*

- *a third line intersecting one of two parallel lines intersects the other one, too.*
- *a third line parallel to one of two parallel lines, is parallel to the other one, too.*

Problem II.4.2. *Explain why Proclus' Lemma is an easy consequence of the uniqueness of parallels. Convince yourself that, conversely, Proclus' Lemma implies the uniqueness of parallels.*

Solution. Suppose towards a contradiction that the transversal t intersects one of the parallel lines l and m, but not the other one.

We may assume that P is the intersection point of lines t and m. If lines t and l would not intersect, then t and m would be two different parallels of line l through point P. This contradicts the uniqueness of parallels.

Conversely, we now assume Proclus' Lemma to be true and check the uniqueness of the parallel to a given line l through a given point P not on line l. Let m and t be two parallels to line l through point P—these may be equal or different lines. The line t is a transversal intersecting one of the two parallel lines $m \parallel l$ at point P. Hence it intersects the second line l, too, contrary to the assumption. The only possibility left is that $m = t$. Hence the parallel to a given line through a given point is unique. □

Remark. Don't fool yourself to think that all these words amount to a proof of the parallel postulate! All we have shown is: <u>either</u> both uniqueness of parallels and Proclus' Lemma hold <u>or</u> neither uniqueness of parallels nor Proclus' Lemma do hold.

Problem II.4.3. *Given any incidence plane where Hilbert's Parallel Postulate for plane geometry II.4.4 holds. Convince yourself that the relation that two lines are either equal or parallel is an equivalence relation.*

Answer. For any two lines, let $l \sim m$ mean $l = m$ or $l \parallel m$. The relation $\sim$ has the three defining properties of an equivalence relation. Indeed this relation is

reflexive: since $l = l$ is a logical axiom.

symmetric: since $l = m$ implies $m = l$, and $l \parallel m$ implies $m \parallel l$.

transitive: assume that $m \sim l$ and $l \sim k$. If two of the three lines are equal, we substitute equals to get $m \sim k$. We now assume that all three lines are different. The two lines m and k are either parallel or they intersect at one point, by Hilbert's first Proposition II.4.1. If m and k would intersect at point P, the line l would have two different parallels through P, which is impossible in an affine plane. Therefore m and k are parallel, as to be shown.

A relation that is reflexive, symmetric and transitive is called an equivalence relation.

Corollary 9 (About uniqueness of parallels). *For any incidence plane the following statements are equivalent:*

(a) *uniqueness of parallels,—stated in Hilbert's Parallel Postulate II.4.4 for plane geometry,—holds;*

(b) *any third line intersecting one of any two parallel lines intersects the other one, too;*

(c) *any third line parallel to one of any two parallel lines, is parallel to the other one, too;*

(d) *being equal or parallel defines an equivalence relation among the lines.*

Definition II.4.8 (Pencil). An equivalence class of (equal or) parallel lines is called a *pencil.*

Definition II.4.9 (Isomorphism of incidence planes). Two models for incidence planes are called *isomorphic* if and only if there exists a bijection ϕ between the points of the two planes, and a bijection φ between the lines of the two planes such that incidence is preserved.

Remark. For incidence to be preserved, the two bijections are required to satisfy the assumption

$$\phi(PQ) = \phi(P)\phi(Q) \tag{II.4.1}$$

for all points P, Q from the first incidence plane. In other words, if the bijection of the points maps $P \mapsto \phi(P)$ and $Q \mapsto \phi(Q)$, then the bijection of the lines maps $PQ \mapsto \phi(P)\phi(Q)$.

II.4.2 Affine Incidence Planes

Definition II.4.10 (Affine plane). An *affine plane* is a set of points, and a set of lines, satisfying the axioms:

A.1 Every two different points lie on exactly one line.

A.2 If point P does not lie on a line l, there exists exactly one line m through the point P that does not intersect l.

A.3 There exist three points that do not lie on a line.

Problem II.4.4. *The reader should convince himself that in an affine plane*

1. *on every line lie at least two points;*

2. *in every point intersect at least two lines.*

Proof. Let A, B, C be the three points not on a line which exist according to axiom (A.3).

1. Given is any line l. At most one of the three lines AB, BC and AC is parallel or equal to the given line l. We may assume that neither line AB nor line AC is parallel or equal to the given line l. Hence they intersect the given line in the points $P = AB \cap l$ and $Q = AC \cap l$.

   - If these two points $P \neq Q$ are different, we are ready.
   - We get a more involved argument in the special case that $P = Q$. In this case, we get $A = P = Q$. If the line l intersects the line BC, we get a second intersection point $R = BC \cap l$ on the line l and are ready again.
   - Still open is only the case that lines BC and l are parallel and the point $A = P$ lies on the line l. Let m be the parallel to line $c = AB$ through point C. Since $m \neq BC \parallel l$, the line m is not parallel to l. Hence these two lines intersect at point $R = m \cap l \neq A$, which is a second point on the line l.

2. Given is any point P. Assume that $P \neq A, B, C$. Since these three points do not lie on a line, we get from lines PA, PB and PC at least two different ones through point P.

   In the special case that $P = A$, we get two different lines AB and AC through point P. We argue similarly in the special cases where $P = B$ or $P = C$.

□

Axiom (A.1) repeats Hilbert's incidence axioms (I.1) and (I.2). Axiom (A.2) is the Euclidean parallel property. In problem II.4.4, it is proved that on every line of an affine plane lie at least two points. Together with (A.3), we see that Hilbert's incidence axiom (I.3) holds for an affine plane, too. Hence an affine plane is just an incidence plane for which the Euclidean parallel property holds.

Corollary 10 (Characterizations of an affine plane). *For any incidence plane the following statements are equivalent:*

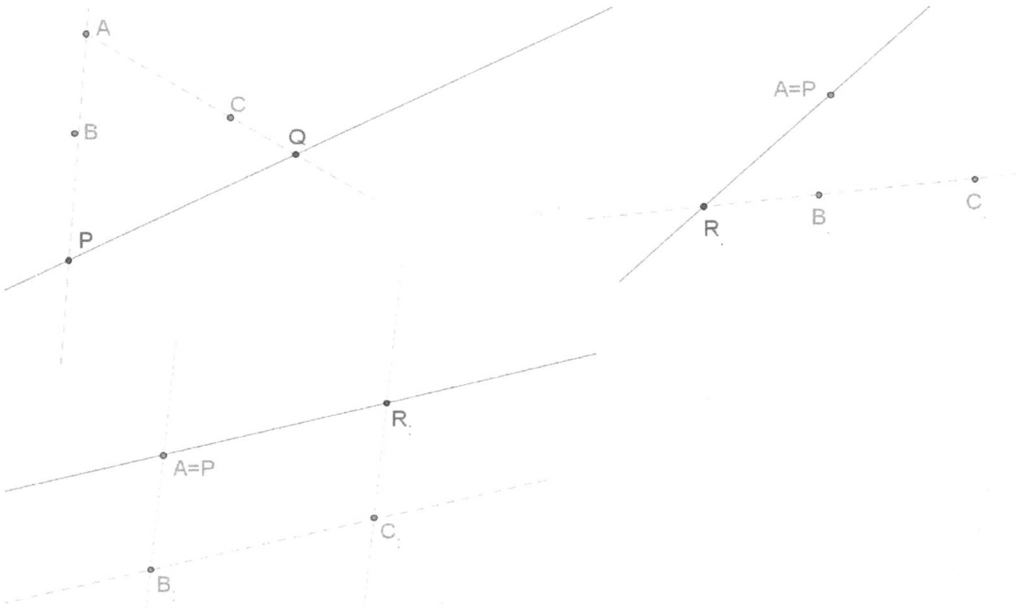

Figure II.4.1: In an affine plane, we find a second point on any given line.

(a) *The incidence plane is an affine plane;*

(b) *for every line l and for every point P lying not on l, there exists a unique parallel m to l through point P. Thus the Euclidean parallel postulate II.4.3 holds.*

II.4.3 Introduction of Coordinates

Definition II.4.11 (The Cartesian plane). The `"points"` of the Cartesian plane are ordered pairs (x, y) of real numbers $x, y \in \mathbf{R}$. The `"lines"` of the Cartesian plane are given by equations

$$ax + by + c = 0 \tag{II.4.2}$$

with reals coefficients a, b, c, and a and b not both equal to zero. A `"point lies on a line"` if and only if the coordinate pair (x, y) satisfies the equation of the line. `Two lines are equal` if and only if they contain the same points.

Remark. Because coefficients (a, b, c) and $(\lambda a, \lambda b, \lambda c)$ with $\lambda \neq 0$ give rise to the same line, the triples (a, b, c) are *homogeneous coordinates.*

Theorem II.4.1. *The Cartesian plane is an affine plane.*

Because the construction of the affine plane from the real numbers uses only addition, subtraction, multiplication and division, replacing the real numbers by any finite or infinite field **F** and doing the same construction once more, leads to an affine plane, too. Thus one gets the following general definition.

Definition II.4.12 (The Cartesian plane over field F). The `"points"` of the Cartesian plane are ordered pairs (x, y) of elements $x, y \in \mathbf{F}$. The `"lines"` of the Cartesian plane of the field **F** are equations

$$ax + by + c = 0$$

with coefficients a, b, c from the field, of which a and b are not both zero. A `"point lies on a line"` if and only if the coordinate pair (x, y) satisfies the equation of the line. `Two lines are equal` if and only if they contain the same points.

Theorem II.4.2 (The Cartesian plane over an arbitrary field). *In a Cartesian plane over any field, there exist a unique line between any two points. There exists a unique parallel to a line through a given point. Hilbert's axioms* (I.1)(I.2)(I.3a)(I.3b) *and the Euclidean parallel postulate* (IV*) *hold. Hence a Cartesian plane over a field is an affine plane.*

Remark. Any axioms may only be used without further justification, in the context that they refer to primary (undefined) elements and relations. The situation is different for the Cartesian plane introduced above, which is another important example for a model or structure. For a *model* or *structure*, the primary elements and relations are replaced by elements and relations which are *explicitly defined.* For a model one has to check whether the *axioms are valid* for these (new) elements and relations.

Thus to check theorem II.4.2, one has to give a justification, why the incidence axioms (I.1), (I,2), (I.3a), (I.3b) and the Euclidean parallel postulate hold for the `"points"`, the `"lines"` and the `"incidences"` of the Cartesian plane specified above.

Problem II.4.5. *Use your knowledge of college algebra, and give in a few words an explanation why Theorem II.4.2 is valid.*

Answer. We begin by checking postulates (I.1) and (I.2). Given are two distinct points (x_1, y_1) and (x_2, y_2) through which we need a unique line. In the usual way, one defines the slope

$$m = \frac{y_2 - y_1}{x_2 - x_1}$$

and puts $m = \infty$ in the special case that $x_1 = x_2$. The line through point (x_1, y_1) with slope m has the equation

$$y = y_1 + m(x - x_1) \tag{II.4.3}$$

in the usual case of finite slope, and $x = x_1 = x_2$ for $m = \infty$. This equation of a line in this form is uniquely determined by two points $(x_1, y_1) \neq (x_2, y_2)$, as the reader should check. Hence two lines through two distinct points contain the same points and are equal.

To check axiom (I.3a), we use that each field contains the two different elements $0 \neq 1$. On any given line with finite slope, we get two points on a line by putting $x = 0, x = 1$. We put $y = 0, y = 1$ for a vertical line. To check axiom (I.3b), we may for example use the three points $(0,0)$, $(1,0)$ and $(1,1)$. Hence we see that the Cartesian plane over any field is an incidence plane.

To get an affine plane, we need still to check that the Euclidean parallel postulate (IV*) does hold. We see that two distinct lines

$$y = mx + k \text{ and } y = m'x + k'$$

with different slopes $m \neq m'$ do intersect. But two distinct lines with the same slope have $m = m'$ and $k \neq k'$,— they do not intersect. Too, one may include vertical line in this statement. Given any line $y = mx + k$ and point (x_1, y_1), the parallel through the point is obtained by the point-slope formula (II.4.3). An easy reasoning shows that Euclidean parallel postulate (IV*) does hold in the case of a vertical line, too.

II.4.4 Propositions for Three-dimensional Incidence Spaces

We go back to Hilbert's system of axioms, and investigate now the three-dimensional incidence spaces. Among the consequences of the axioms of incidence, Hilbert spells out two propositions.

Proposition II.4.1 (Hilbert's Proposition 1).

(a) *Any two different lines have either no or exactly one point in common.*

(b) *Any two different planes have either no point in common, or they have one line and no further points in common.*

(c) *A plane and a line not lying in this plane have either no or exactly one point in common.*

Problem II.4.6. *Prove statement* (b). *Use definite names for the objects, explain the steps, and state which axioms are involved. Prove statement* (c). *Write a short clear proof.*

Answer. **(b)** Take any two different planes $\alpha \neq \beta$. If they do not intersect, we are ready. Assume that the two planes have point of intersection A in common. By axiom (I.7), the two planes have a further point $B \neq A$ in common. By axioms (I.1) and (I.2), there exists a unique line a through points A and B. By axiom (I.6), every point of line a lies in the plane α.

Reasoning about the same points A and B and line a, axiom (I.6) yields that every point of line a lies in the plane β, too. In other words, we get the inclusion

$$a \subseteq \alpha \cap \beta$$

We need still to check that the two planes α and β do not have any point outside the line a in common.

Take any point C lying in both planes α and β. We assume towards a contradiction that point C does not lie on the line a. Both planes α and β contain the three points A, B and C. By axiom (I.5), there exists at most one plane through the three points A, B and C. Hence these two planes α and β would be equal, contradicting the assumption $\alpha \neq \beta$.

Hence the two planes α and β have no point outside the line a in common. Together we get the equality

$$a = \alpha \cap \beta$$

We have shown that two planes intersecting at one point have a line through this point as their intersection.

(c) Take a plane α and a line a not lying in α. We assume towards a contradiction they intersect at two different points $A \neq B$. By axiom (I.6), every point of a lies in the plane α, contradicting the assumption that the line a is not lying in α.

Hence the plane α and line a have at most one point in common.

Proposition II.4.2 (Hilbert's Proposition 2).

(a) *Given a line and a point not lying on it, there exists a unique plane containing the line and the point.*

(b) *Given two different intersecting lines, there exists a unique plane containing these two lines.*

As far as two dimensional geometry is concerned, Proposition II.4.1 reduces to the one simple statement that any two different lines either intersect at one point, or are parallel. But the present section is concerned with the three-dimensional case.

Proof. **(a)** Given is the line a and a point C not lying on it. By axiom (I.3a), there exist two points $A \neq B$ on the line a. The three points A, B, C do not lie on a line. Hence by axioms (I.4) and (I.5), there exists a unique plane α through these three points. By axiom (I.6), all points of line a lie in the plane α. Hence this plane contains both the given line and the given point. Uniqueness of the plane follows by axiom (I.5).

(b) Given are two different lines a and b, intersecting in point C. By axiom (I.3a), there exists a second point $A \neq C$ on line a, and a point $B \neq C$ on the line b. The three points A, B and C do not lie on one line, since the given lines $a \neq b$ are different.

Hence, again by axioms (I.4) and (I.5), there exists a unique plane α through these three points. By axiom (I.6), all points of line a, as well as of line b lie in the plane α. Hence this plane contains both given lines. Uniqueness of the plane follows by axiom (I.5).

□

Lemma II.4.4. *Any three lines in three-dimensional incidence space that go through one point but do not lie in a plane are different from each other.*

Proof. Let the three lines g, h, l all go through point P but not lie in a plane. The assumption that two of them are equal leads to a contradiction. Indeed, under the assumption that two of them would be the same line $g = h$, Hilbert's proposition II.4.2 part (b) would imply existence of a plane through line g and the third line l, since they intersect at point P. Hence all three lines would lie in a plane, a contradiction. Because of this contradiction, we see all three lines are different from each other. □

Proposition II.4.3 (Three lines through any point). *Through any point of a three-dimensional incidence space, there exist three lines not lying in a plane. Especially, these three lines are different from each other.*

A second variant. We use a modification of the proof for lemma II.4.2. Let any point P be given. By axiom (I.8), there exist at least four points not lying in a plane. We call them A, B, C and D.

In the case that point P is among the four points A, B, C, D, we may assume $P = A$. We draw the three lines AB, AC and AD. We check that these three lines do not lie in a plane,—otherwise the four points A, B, C, D would lie in a plane, contrary to the assumption.

In the case that point P is different from all four points A, B, C, D, we draw all four lines PA, PB, PC and PD.

- In the case that either lines PA, PB, PC or lines PA, PB, PD do not lie in a plane, we are ready.
- It may happen that both lines PA, PB, PC lie in the plane γ, and lines PA, PB, PC lie in a plane δ.

  But it cannot happen that any three lines, say e.g. PA, PB, PC, are equal. Otherwise Hilbert's proposition II.4.2 part (a) would imply existence of a plane through line $l = PA = PB = PC$ and the fourth point D. Thus the four points A, B, C, D would lie in a plane, contrary to the assumption.

  Hence lines PA, PB, PC determine the plane γ uniquely. Too, the lines PA, PB, PD determine the plane δ uniquely. Moreover, these are two different planes,—otherwise all four lines PA, PB, PC, PD and all the four points A, B, C, D would lie in a plane, contrary to the assumption.

  Now Hilbert's proposition II.4.1 part (b) implies that the intersection $\gamma \cap \delta$ is a unique line, which is line $PA = PB$.

  We conclude that the lines PB, PC, PD cannot lie in a plane,—otherwise all four lines PA, PB, PC, PD and all the four points A, B, C, D would lie in a plane, contrary to the assumption. Hence we have obtained three lines PB, PC and PD not lying in a plane.

In all cases we have obtained three different lines through the arbitrary point P.

□

Proposition II.4.4 (Every plane is spanned by three points). *In every plane, there exist three points not lying on a line. Hence any plane can be uniquely specified by three points lying on it.*

Proof. By axiom (I.4a), there exists a point P in the given plane π. By proposition II.4.3, there exist three lines g, h and l through point P, which do not lie in a plane.

By Hilbert's proposition II.4.2 part (b), there exists a plane spanned by the lines g and h, which we denote by $\alpha := \text{span}(P, g, h)$. Similarly, there exist the planes $\beta := \text{span}(P, g, l)$ and $\gamma := \text{span}(P, h, l)$. All three planes are different from each other,—otherwise the three lines g, h, l would lie in a plane.

By Hilbert's Proposition II.4.1 part (b), one gets the intersection lines $\alpha \cap \pi$, $\beta \cap \pi$ and $\gamma \cap \pi$. Two of them may be equal. Nevertheless, it is impossible that they are all three equal,—otherwise all three line g, h, and l would be equal.

We may assume the two lines $\alpha \cap \pi \neq \beta \cap \pi$ to be different. By axiom (I.3a), a second point $Q \neq P$ on the line $\alpha \cap \pi$ exists. Similarly, a point $R \neq P$ exists on the second line $\beta \cap \pi$, again by axiom (I.3a). All three points P, Q, R lie in the given plane π, but not on one line. By axiom (I.5), they determine the plane π uniquely. Hence we say they <u>span</u> the plane π. □

Part III

From Axioms to Coordinates, and Back

III.5 Arithmetic of Segments: Hilbert's Road from Geometry to Algebra

In this section, we explain Hilbert's procedure to construct an arithmetic of segments, also called *Streckenrechnung*. Hilbert constructs a field from a given model of geometry and thus opens up a general road from geometry to algebra.

The approach is explained in Hilbert's foundations. It works for an arbitrary affine plane for which the theorem of Pappus holds. Since existence and uniqueness of the parallel is assumed, one can take advantage of similar triangles in the construction of the field of segment arithmetic. Neither one of the axioms of continuity shall be assumed. Depending on the given model of geometry, we obtain a variety of different fields, rather than the real numbers exclusively.

One can even generalize the approach, and assume only the weaker theorem of Desargues instead of Pappus. In this case, one obtains on the side of algebra only a skew field—thus only the associative law still holds, but the commutative law does no longer hold. These important results are stated once more in the Main Theorems of Desargues and of Pappus. In the context of projective geometry (see volume VI), the commutative law turns out to be equivalent to the theorem of Pappus and the associative law to the theorem of Desargues.

As an alternative, we may avoid the theorem of Pappus and get the segment arithmetic more directly from angles in a circle via Euclid's III.21. This approach is used in Hartshorne's book and presented here in an own variant. It works in every Pythagorean plane and breaks any connection with ideas from projective geometry. Recall the definitions:

Definition (II.3.2 Pythagorean plane). A *Pythagorean plane* is a Hilbert plane for which the Euclidean parallel postulate holds.

Definition III.5.1 (Pythagorean field). A *Pythagorean field* is an ordered field with the property

(Pyth) If $a, b \in \mathbf{F}$, then $\sqrt{a^2 + b^2} \in \mathbf{F}$.

Definition III.5.2 (Pappus' configuration). We call the system of two distinct intersecting or parallel lines, with three points on each one of them, and the M-W hexagon zig-zaging between those triplets the *configuration of Pappus*.

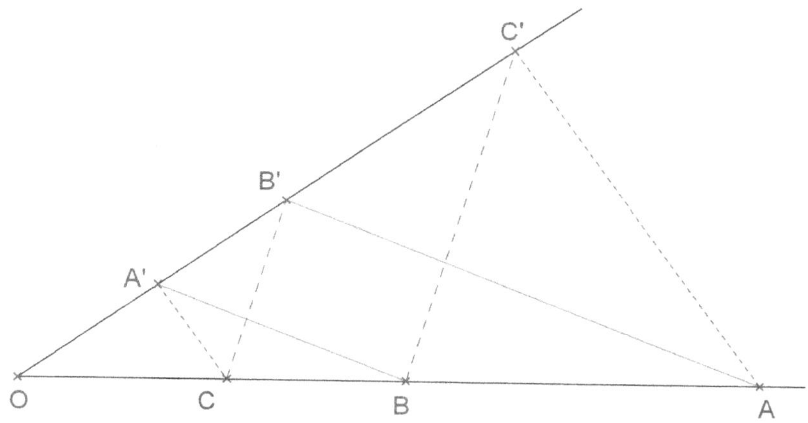

Figure III.5.1: For Pappus' configuration: If $BC' \parallel B'C$ and $AC' \parallel A'C$, then $AB' \parallel A'B$.

In the present figure, we have points A, B, C on the horizontal line, three points A', B', C' on the tilted line. In the hexagon $AB'CA'BC'$ there are three pairs of opposite sides: the three pairs of lines

$$\begin{array}{ccc} A'B & \text{and} & AB' \\ B'C & \text{and} & BC' \\ C'A & \text{and} & CA' \end{array}$$

the parallelism of which has to be investigated. Thus we are lead to conjecture that the hexagon of Pappus' configuration containing two pairs of opposite parallel sides shall have a third pair of opposite parallel sides, too. A formulation with definitely specified variables is given here.

Theorem III.5.1 (Pappus Theorem in parallel setting). *We let A, B, C and A', B', C' be both three points on two intersecting lines, all different from the intersection point. If the lines BC' and $B'C$ are parallel, and the lines AC' and $A'C$ are parallel, then the lines AB' and $A'B$ and parallel, too.*

Remark. In Hilbert's foundations [10], this theorem is named after Pascal. Pascal's name is now generally associated with the theorem about the hexagon in a circle or conic section. Following Stillwell's book [18]. I prefer to use the name Pappus for the theorem which does not involve any circle or conic section.

Main Theorem 1. *Any affine plane for which the Theorem of Pappus holds is isomorphic to the Cartesian plane over a field. Conversely, the Theorem of Pappus holds in a Cartesian plane over a field.*

The first result is elaborated in this section in some detail. The second statement repeats part of the Theorem of Pappus, and can be checked directly with some rather tedious calculations which we leave out. Furthermore, this section gives the detailed direct proof of the following result:

Theorem III.5.2. *For any Pythagorean plane the field* $\mathbf{F}$ *of segment lengths is a Pythagorean field. Any Pythagorean plane is isomorphic to the Cartesian plane* $\mathbf{F}^2$ *over its field* $\mathbf{F}$ *of segment lengths.*

III.5.1 Fields

Definition III.5.3 (Field). A *field* is a system of undefined elements (symbols, numbers), two binary relation symbols $+$ and $\cdot$ and two constants 0 and 1 with the following properties:

(1) (i) For any two numbers a and b exists the *sum* $a + b$.

- **(ii)** $(a + b) + c = a + (b + c)$ for any a, b, c.
- **(iii)** $a + b = b + a$ for any a, b.
- **(iv)** $a + 0 = a$ for any a.
- **(v)** for each number a there exists a number $-a$ such that $a + (-a) = 0$.

(2) (i) For any two numbers a and b exists the *product* $a \cdot b$.

- **(ii)** $(a \cdot b) \cdot c = a \cdot (b \cdot c)$ for any a, b, c.
- **(iv)** $a \cdot 1 = 1 \cdot a = a$ for any a.
- **(v)** for each number $a \neq 0$ there exists a number a^{-1} such that $a \cdot a^{-1} = a^{-1} \cdot a = 1$.

(3) (i) $0 \neq 1$.

- **(ii)** $(a + b) \cdot c = a \cdot c + b \cdot c$ for any a, b, c.
- **(iii)** $c \cdot (a + b) = c \cdot a + c \cdot b$ for any a, b, c.

(4) $a \cdot b = b \cdot a$ for any a, b.

A system of elements satisfying only (1)(2)(3) is called a *skew field.*

Any model for the field axioms is simply called a field, too. In this context, we denote by $\mathbf{F}$ the set of elements. Clearly $0, 1 \in \mathbf{F}$.

III.5.2 Construction of the Field of Segment Arithmetic

To relate Euclidean geometry to algebra, we need to construct a field from the congruence classes of segments.

One can define the sum of any two segments by putting them together along a line. Especially, if point B lies between A and C, we define $AC = AB + BC$. More generally, the sum of any two segments is defined by Definition V.12.1. Because of Hilbert's postulate III.3, the addition of segments defines even an addition for the congruence classes of segments. As already shown in Proposition V.12.4 in the section about congruence of segments, angles and triangles, we get an ordered Abelian group from these congruence classes, their negatives and the zero element.

In the theorem 5 about the measurement of segments, we have already used a *unit segment*. To this unit segment, say OE, we assign the length 1. The construction of the length for any segment as an (finite or infinite) binary fraction done in the measurement theorem 5 depends on the Archimedean axiom.

Taking the Archimedean axiom for granted, we could simply define the product and the quotient of segment via the product and quotient of the corresponding binary fractions. Although this "old segment arithmetic" is very practical (and usually tacitly assumed in high school), it depends essentially on the Archimedean axiom, and the possibility to approximate irrational numbers by rational numbers. Only to <u>some</u> very special segments, a finite binary or proper fraction can be assigned as length. Hence the old segment arithmetic is not suitable for a general constructive definition of the multiplication of the field.

The essential step for the "new segment arithmetic" claimed by Theorem III.5.2 is the correct definition of the product of two segments. This is done using the legs of equiangular right triangles. Thus all dependence on the axioms of continuity can be avoided.

Remark. Hilbert even shows in detail how the assumption of any axioms of congruence can be avoided, if instead the theorem of Desargues is assumed. We can leave these considerations aside, and define the sum of segments still via Definition V.12.1 from neutral geometry.

To be remembered. For the definition of the product ab, we use two similar right triangles. The first triangle has the legs the <u>first factor</u> a and 1, the second one has as its legs ab and the <u>second factor</u> b. □

Look at the figure on page 101. On the perpendicular ray through O, we have put the segment with prescribed length

$$|OA'| = a$$

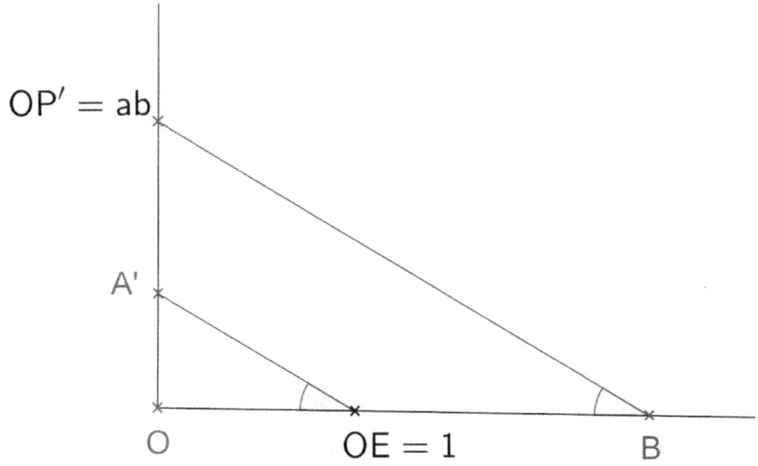

Figure III.5.2: Construction of the product ab for given segments $|OA'| = a, |OB| = b$ and $|OE| = 1$.

On the horizontal ray emanating from vertex O to the right, there are put points O, E and B such that the segments measured from the origin O have the prescribed lengths:

$$|OE| = 1\,,\ |OB| = b$$

Now we draw the parallel to line EA' through point B. Why does this parallel does intersect the vertical line? Indeed, by Proclus' Lemma V.12.2, a line intersecting one of two parallel lines intersects the other one, too. Since the vertical line intersects line EA', it intersects its parallel through B, too. Hence we get the unique intersection point P' and define the product ab as length of the segment OP':

$$ab := |OP'|$$

Indeed, we have used the similar right triangles $\triangle OEA'$ and $\triangle OBP'$.

Question. How does the order of the factors a and b become relevant for the construction?

Answer. The first factor $a = |OA'|$ and the unit segment $|OE| = 1$ appear *together as legs of one triangle* $\triangle OEA'$. The second factor b determine the segment OB' through one end of which one draws the parallels $EA' \parallel BP'$.

Indeed, the product of segments a and b turns out to have the property that the rectangle with sides a and b has the same area as the rectangle with sides 1 and ab.

Remark. The definition of area does not automatically imply commutativity. On the contrary, the usual simple definition of the area of the rectangle as the product of its sides is only possible *because of commutativity.* [8]

The first obvious question is whether our definition given above makes the product commutative. Furthermore, we need to go on and check whether the remaining axioms required for a field are satisfied, and whether we get an ordered field.

Commutative law

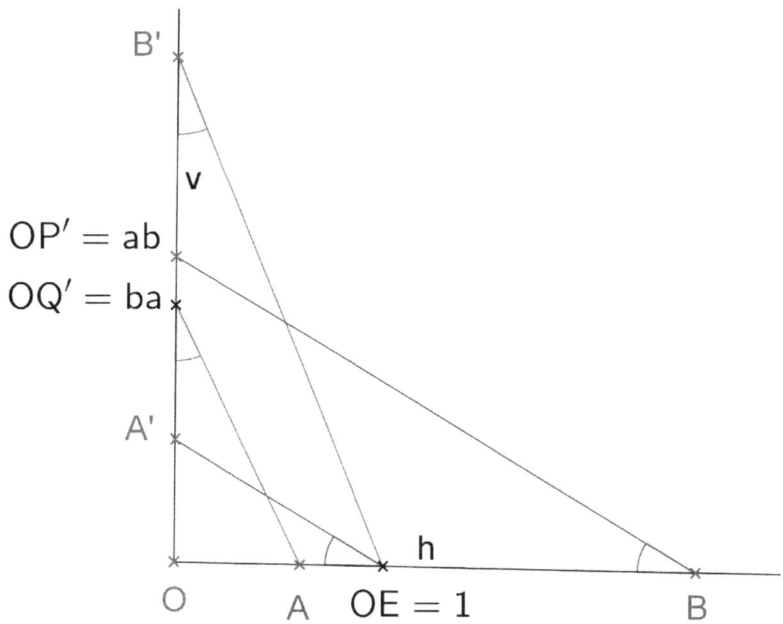

Figure III.5.3: Construction of the products ab and ba, with an empirical error exaggerated.

To check the commutative law, we repeat the above construction, with the roles of a and b exchanged. We construct point A on the horizontal line such that

$$|OA| = a$$

and a point on the perpendicular line such that

$$|OB'| = b$$

[8] It would be interesting to find out in a psychological study who really knows that!

Now we draw the parallel to line EB' through point A, and get the intersection point Q'. By definition ba is length of the segment OQ':

$$ba = |OQ'|$$

As you can see from the figure on page 102, it is now really a question from geometry, whether $ab = ba$!

Proposition III.5.1. *The segment arithmetic is commutative in any affine plane where the Theorem of Pappus holds.*

Proof. The product ab is constructed as in the figure on page 102, by the pair of parallels

$$EA' \parallel BP'$$

Because of the isosceles and equiangular triangles $\triangle OAA'$ and OBB', we get another pair of parallel lines $AA' \parallel BB'$. The hexagon $AP'BB'EA'$ and the horizontal and vertical lines h and v are a Pappus configuration.

Remark. To get the three pairs of opposite sides of the hexagon, it is convenient to shift the points cyclically by three:

$$\begin{array}{cccccc} A & P' & B & B' & E & A' \\ B' & E & A' & A & P' & B \end{array}$$

The three pairs of opposite sides of the hexagon appear underneath each other:

$$\begin{array}{ccc} AP' & P'B & BB' \\ B'E & EA' & A'A \end{array}$$

The Pappus configuration contains the two pairs of parallel opposite sides

$$EA' \parallel BP'$$
$$AA' \parallel BB'$$

Hence, by the theorem of Pappus, there is a third pair of parallels

$$EB' \parallel AP'$$

The product ba is constructed as in the figure on page 102, by the pair of parallels

$$EB' \parallel AQ'$$

Both $EB' \parallel AQ'$ and $EB' \parallel AP'$, and both P' and Q' lie on the vertical line v. The uniqueness of the parallels and the uniqueness of intersection point implies $P' = Q'$. Hence $ab = |OP'| = |OQ'| = ba$, as to be shown. □

Remark. Conversely, the commutative law of multiplication, together with the theorem that equiangular triangles have proportional sides, can be used to get Pappus theorem in the simplified version.

Remark. In this and the following proofs, we consider only the case of positive quantities a, b, c. It is left to the reader to check that the corresponding results hold for positive as well as negative quantities.

The distributive law

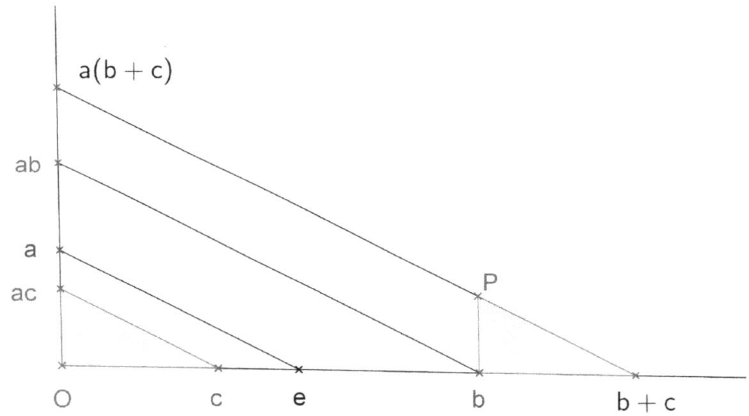

Figure III.5.4: Proof of the distributive law $ab + ac = a(b + c)$.

Proposition III.5.2. *The segment arithmetic is distributive in any Pythagorean plane.*

Proof. Given are the segment lengths a, b, c, which we may assume to be positive. We use two perpendicular rays with the common origin O, and transfer the segment a onto one of them,—the vertical one in the drawing; and the two other segments b and c and the unit segment $[O, e]$ onto the other one,—the horizontal one in the drawing.

Next we construct the segment $[b, b + c]$, by transferring the segment $[O, c]$ onto the ray with vertex b on the horizontal axis in order to get segment $[O, b + c]$. The construction of the products ab, ac, and $a(b + c)$ is shown in the figure on page 104. In all three cases, we use parallels to the line EA', where A' is the point on the vertical axis with $a = |OA'|$. These parallels go through points b, c and $(b+c)$ on the horizontal axis, respectively. On the vertical axis, we get the intersection points ab, ac and $a(b + c)$, the distance of which to the origin define the respective products.

Finally, we draw the parallel to the vertical axis through point b, and get intersection point P with the line through points $b+c$ and $a(b+c)$. By ASA congruence, the triangles $\triangle Ob(ab)$ and $\triangle b(b+c)P$ are congruent, since they are equiangular and have one pair of congruent sides. Hence the segments $[O, ac]$ and $[b, P]$ are congruent. Furthermore, the segments $[b, P]$ and $[ab, a(b+c)]$ are congruent since they are opposite sides of a parallelogram. Hence $[O, ac] \cong [ab, a(b+c)]$.

By definition of segment addition, which is now done on the vertical axis, we know that $[O, ac] \cong [ab, ab+ac]$. Hence we finally get $ab+ac = a(b+c)$, as to be shown. □

The associative law

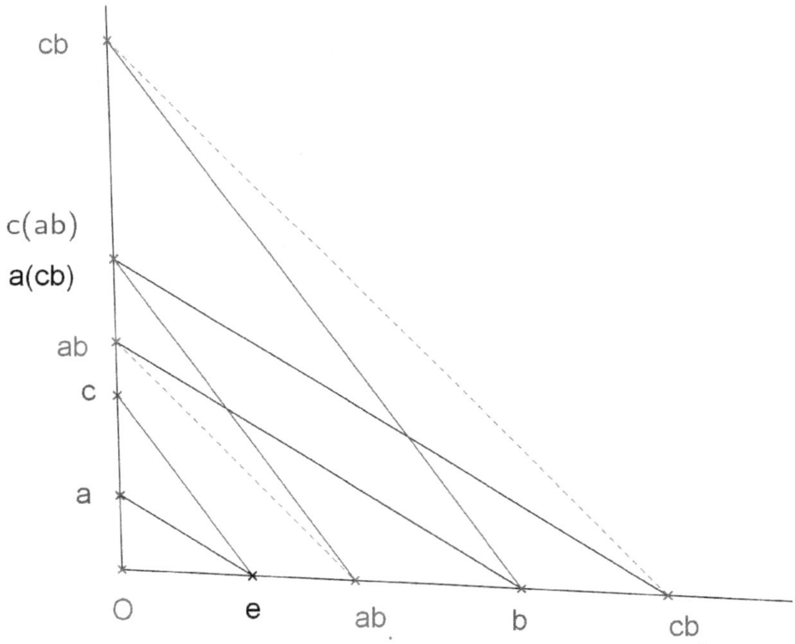

Figure III.5.5: The proof of the associative law yields at first $c(ab) = a(cb)$.

Proposition III.5.3. *The segment arithmetic is associative in any affine plane where the Theorem of Pappus holds.*

Proof. From given segments a and c on the vertical line, and $e = 1$ and b on the horizontal line, we construct at first segments ab and cb on the vertical line. These two segments are transferred onto the horizontal axis, using 45°-lines. Finally, one gets, both on the vertical line, segment $c(ab)$ from the parallels $[e, c] \parallel [ab, c(ab)]$ and segments $a(cb)$ from the parallels $[e, a] \parallel [cb, a(cb)]$.

The three points ab, b and cb on the horizontal axis and the three points ab, $c(ab)$ and cb on the vertical axis give a Pappus configuration with two, and hence three pairs of parallels. We conclude that $a(cb) = c(ab)$. By applying the already proved commutative law twice, we get

$$a(bc) = a(cb) = c(ab) = (ab)c$$

□

The circum-center of a triangle will be used in the next proof. For completeness and convenience of the reader, I include the following paragraph.

The circum-center

Lemma III.5.1. *In a Pythagorean plane are given any angle $\angle(h, k)$ with vertex O, and two lines p and q which intersect the sides of the angle at points $P = h \cap p$ and $Q = k \cap q$. We assume that $P \neq O$ and $Q \neq O$. If $p \perp$ line h and $q \perp$ line k, then the lines p and q intersect at a unique point.*

Proof. Let $O = h \cap k$ be the vertex of angle $\angle(h, k)$. Let $P = h \cap p$ and $Q = k \cap q$ be the points where the perpendiculars are erected on the sides of the angle. We prove:

"$p \neq q$": By assumption $O \neq P$ and $O \neq Q$ and hence triangle $\triangle OPQ$ exists. The case $p = q$ leads to a contradiction as follows. We would obtain a triangle $\triangle OPQ$ with two right angles at vertices P and Q. By the exterior angle theorem V.12.7 (indeed already the basic Lemma given before) such a triangle cannot exist.

"$p \nparallel q$": We know already that $p \neq q$. Because of the assumptions $P \neq O$ and $Q \neq O$, neither lines $p \perp \overrightarrow{OP}$ nor $q \perp \overrightarrow{OQ}$ can go through the vertex O. We show that the assumption $p \parallel q$ leads to a contradiction as follows. With assumption $p \parallel q$ we may use Proclus' lemma V.12.2. Because the line of h intersects p, Proclus' lemma yields that the line of h intersects the parallel $q \parallel p$, too.

Let $R =$ line $h \cap q$ be their intersection point. Since line q does not go through point O, we conclude $R \neq O$. Hence RO is the line of ray h. Since $Q \neq O$ has been assumed, we conclude that QO is the line of ray k. Since $\angle(h, k)$ is

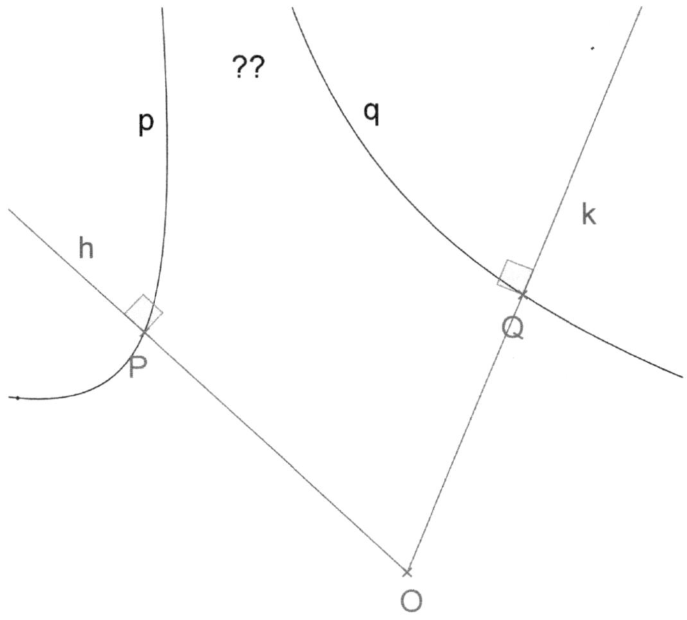

Figure III.5.6: Do the perpendiculars p and q intersect?

assumed to be an angle, the points O, Q and R do not lie on a line. Hence the triangle $\triangle OQR$ exists.

It has been shown in proposition V.12.23 that in a Pythagorean plane, any two parallel lines form congruent z-angles with any transversal.

The line of ray h is transversal to the parallels $p \parallel q$. Too, the line of h is perpendicular to p by assumption. Hence the line of h is perpendicular to q.

By assumption, the line of k is perpendicular to q, too. We have obtained a triangle $\triangle OQR$ with two right angles at vertices Q and R. By the exterior angle theorem such a triangle cannot exist. By this contradiction, the lines p and q cannot be parallel.

The lines p and q are neither equal nor parallel. Hence the lines p and q intersect at a unique point, as to be shown. □

Theorem III.5.3 (The Circum-center). *In a Pythagorean plane, the bisectors of all three sides of any triangle intersect at one point,—which is the center of the circum-circle.*

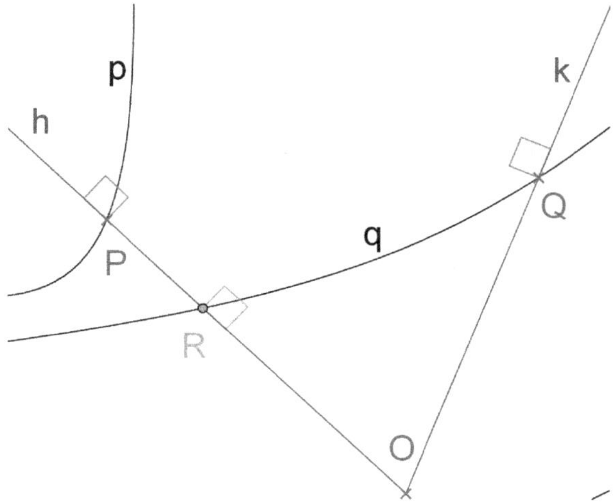

Figure III.5.7: No, the triangle $\triangle OQR$ cannot exist.

Proof. Because of lemma III.5.1, any two perpendicular side-bisectors of the triangle $\triangle ABC$ intersect. Let the bisector p of side AB intersect the bisector q of side BC at the point O. By proposition V.12.16 a point lies on the perpendicular bisector a segment if and only if its distances to the endpoints of the segment are congruent. Since point O lies on the bisector of AB, we conclude $AO \cong BO$. Since point O lies on the bisector of BC, we conclude $BO \cong CO$. Hence point O has congruent distances to all three vertices of triangle $\triangle ABC$. Hence point O is the center of the circum circle. Too, we conclude that O lies on the bisector of the third side CA. □

A direct proof of commutativity of segment arithmetic

Proposition III.5.4. *In any Pythagorean plane, the segment arithmetic is commutative.*

Remark. The following proof of commutativity for the segment arithmetic does not presuppose Pappus' theorem. As a prize, we have to restrict our attention to a Pythagorean plane. We use Euclid III.21 about angles in a circle directly. Thus we get a proof that commutativity of segment arithmetic holds in every Pythagorean plane.

Proof. Given are the positive lengths $a > 0$ and $b > 0$, and a unit segment OE. We take the line OE as horizontal axis, the ray $\overrightarrow{OE}$ as the positive x-axis. At point O the perpendicular is erected to obtain the vertical axis.

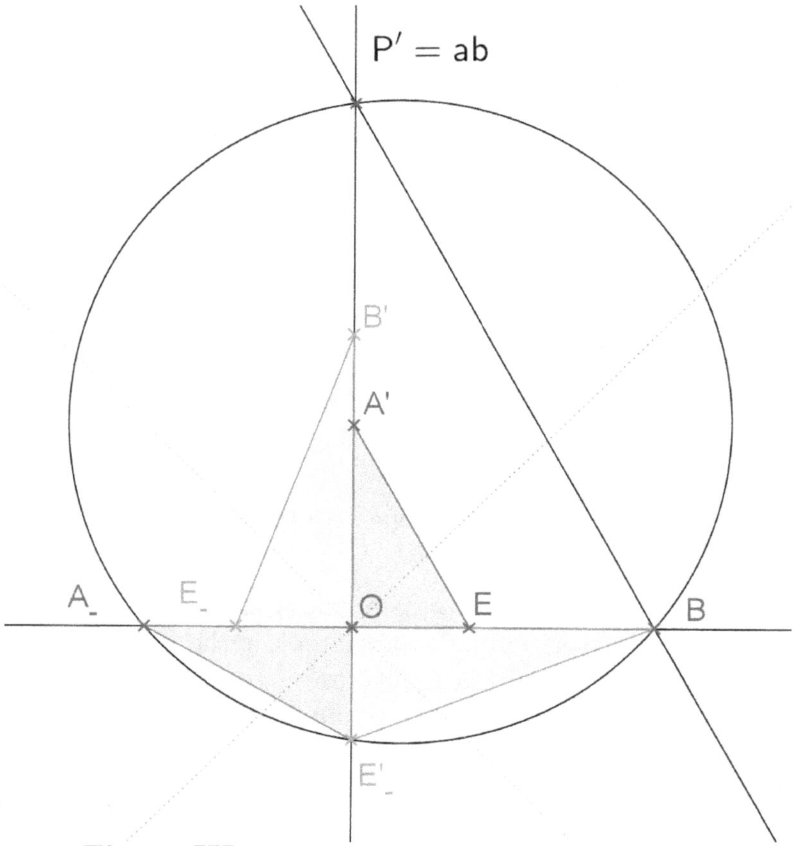

Figure III.5.8: Prove commutativity $ab = ba$.

The reader should look at the figure on page 109 to see how the three points A_-, E'_- and B are obtained: The given positive segment a is transferred to the vertical axis to obtain a point A' such that $|OA'| = a$. The same segment a is transferred to the negative horizontal axis to obtain a point A_- such that $A_- * O * E$ and $|OA_-| = a$. The unit segment is transferred to the negative vertical axis to obtain a point E'_- such that $E'_- * O * A'$ and $|OE'_-| = 1$. The segment b is transferred to the positive horizontal axis to obtain a point B such that $|OB| = b$. The three points A_-, E'_- and B do not lie on a line. By theorem III.5.3, in a Pythagorean plane every triangle has a circum-circle. Hence there *exists a circle* through points A_-, E'_- and B. Indeed, this is the only circle used for the proof.

The origin of coordinates O lies in the segment A_-B and hence in the interior of this circle. (Clearly point O need not be the center of the circle.) By construction, the vertical axis intersects the circle at point E'_- and goes through the point O in the interior of the circle. Hence it is easy to prove that the vertical axis intersects the circle at second point P' for which $E'_- * O * P'$. Note that no circle-line postulate is needed to get point P'.

Question. Recall how the product ab has been constructed.

Answer. As explained by the figure 101, the construction of the product ab uses two similar right triangles. The first triangle has the legs the <u>first factor</u> a and 1, the second one has as its legs ab and the <u>second factor</u> b. To get these triangles, the given positive segment a has been transferred to the positive vertical axis to obtain a point A' such that $|OA'| = a$. To obtain the product ab, we draw the parallel to line EA' through point B and get its intersection point Y with the vertical axis. The product is given by the length $|OY| = ab$ of the segment from the origin.

Question. Why is the parallel obtained for free, — and indeed $Y = P'$?

Answer. We reflect the triangle $\triangle OA'E$ across the $-45°$ line to produce the triangle $\triangle OA_-E'_-$. The triangles $\triangle OA'E \cong \triangle OA_-E'_-$ are congruent, as one easily checks by SAS-congruence. Hence the angles $\angle OA'E \cong \angle OA_-E'_-$ are congruent, too.

The angles $\angle BA_-E'_- \cong \angle BP'E'_-$ are both circumference angles of the arc E'_-B, and both lie on the same side of the chord E'_-B. Hence they are congruent by Euclid III.21. Both steps together show the congruence

$$\angle OA'E \cong \angle OA_-E'_- = \angle BA_-E'_- \cong \angle BP'E'_- = \angle OP'B$$

By Euclid I.27, "congruent z-angles imply parallels" (See proposition V.12.15). In a Pythagorean plane, the parallel is *unique*. Hence the line $P'B$ is the *unique* parallel to line EA' through point B, which is to be used in the construction of the product ab. The intersection point Y with the vertical axis is equal to point P'.

Question. Explain how the product ba has to be constructed, in the left upper quadrant.

Answer. The triangle $\triangle OBE'_-$ is reflected across the $+45°$ line. We obtain the triangle $OB'E_-$, needed to construct the product ba. We draw the line through A'_- parallel to E_-B', and get its intersection point with the vertical axis at point Z with distance $|OZ| = ba$ from the origin.

Question. Explain why, because of these constructions, there appear three congruent angles β.

Answer. The angles $\angle OB'E_- \cong \angle OZA_-$ are congruent because the vertical axis traverses a pair of parallel lines. The angles $\angle OBE'_- \cong \angle OB'E_-$ are congruent because of SAS congruence of the corresponding triangles. We have marked the three congruent angles by β.

Question. Explain why the circle through A_-, E'_- and B—already drawn—contains the point Z.

Answer. The angles $\angle E'_- BA_- \cong \angle E'_- P'A_-$ are circumference angles of the same arc, and lie both on the same side of its chord $E'_- A_-$. Hence they are congruent. From

$$\angle E'_- P'A_- \cong \angle E'_- ZA_-$$

we see that point $P' = Z$ lies on this circle.

Since the circle has only one intersection point P' with the vertical ray $\overrightarrow{OA}$, we conclude that $Y = Z = P'$. Hence we have obtained that both ab and ba is this intersection point, we conclude that $ab = ba$. □

Remark. In the drawing from a former version—which was not from Leibniz' "best of all worlds"—the construction did not come out right. In such a case, would Leibniz have trusted the proof or the drawing?

As an additional benefit, we state:

Corollary 11. *The Theorem of Pappus holds in every Pythagorean plane.*

Alternative proof of associativity

The given positive segments $|OA| = a > 0$ and $|OB| = b > 0$ are transferred to the vertical axis. The segments $|OE| = 1$ and $|OC| = c > 0$ are transferred to the horizontal axis. The products ac and bc are constructed on the vertical axis.

Problem III.5.1. *One goes on, almost as above, with E, A, B replaced by C, AC and BC. The angles α and β each appear four time. Explain in detail how the figure on page 113 yields a proof of $a(bc) = b(ac)$ with any three positive $a, b, c > 0$, and thus confirms the associative law.*

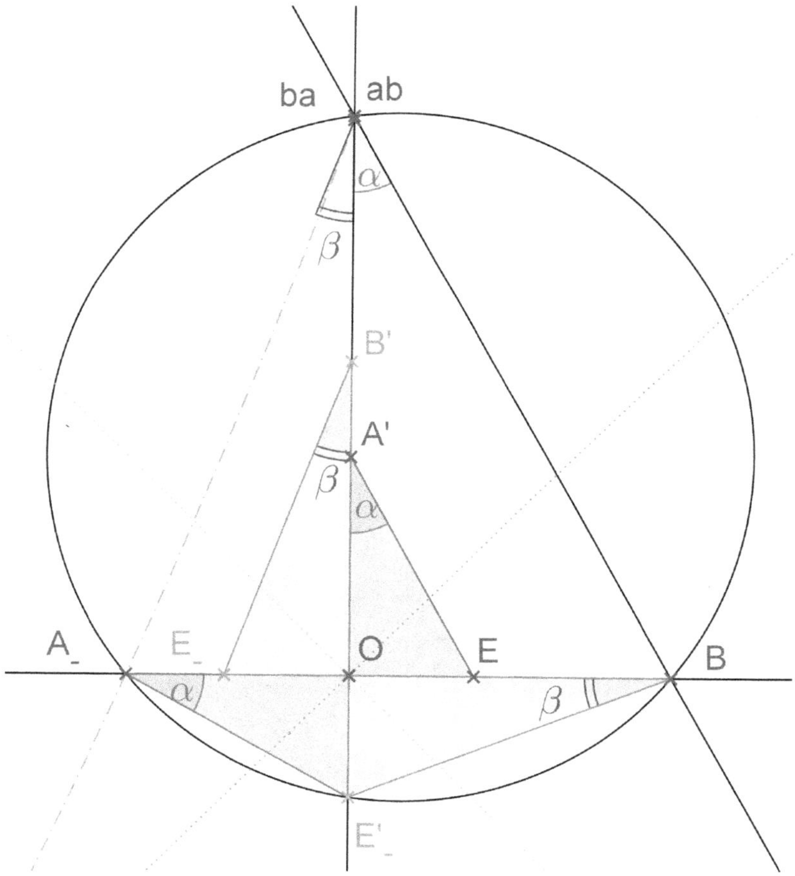

Figure III.5.9: Commutativity $ab = ba$ follows from both lying on the circle.

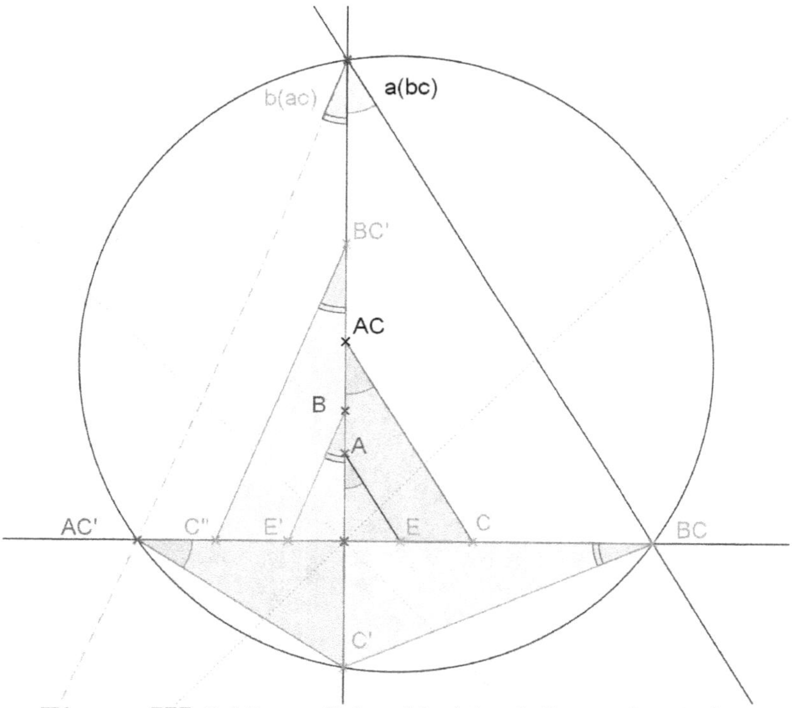

Figure III.5.10: $a(bc) = b(ac)$ both lie on the circle.

III.6 Coordinates are Descartes' Road from Algebra to Geometry

The Cartesian plane over a field **F** takes as coordinates the elements of an arbitrary finite or infinite field **F**— in place of the real number coordinates for the real Cartesian plane. We recall from definition II.4.12:

- The `"points"` of the Cartesian plane are ordered pairs (x, y) of elements $x, y \in$ **F**.
- The `"lines"` of the Cartesian plane of the field **F** are equations

$$ax + by + c = 0$$

  with coefficients a, b, c from the field, of which a and b are not both zero.
- A `"point lies on a line"` if and only if the coordinate pair (x, y) satisfies the equation of the line.

One defines the slope of a line in the usual way. Lines are parallel if and only if they have the same slope. To find the parallel to a given line through a given point, one uses the point slope equation of a straight line. This procedure shows that the Euclidean parallel property holds. Leaving the details aside, we repeat the results:

(**Theorem II.4.2**). *In a Cartesian plane over any field, there exists a unique line between any two points. There exists a unique parallel to a line through a given point. Hilbert's axioms* (I.1)(I.2)(I.3a)(I.3b) *and the Euclidean parallel postulate hold. Hence a Cartesian plane over a field is an affine plane.*

The Theorem of Pappus implies the converse of the elementary Theorem II.4.2. Together, there is the sharp result:

Theorem III.6.1. *Any affine plane is isomorphic to a plane over a field if and only if the Theorem of Pappus holds in its projective completion.*

III.6.1 A Hierarchy of Cartesian Planes

Definition III.6.1 (Ordered field). An *ordered field* is a field with an order relation " $<$ " having the following properties:

(i) For any two numbers a and b, one of the three relations holds: either $a < b$ or $b < a$ or $a = b$.

(ii) For any two numbers a and b, no more than one of the three relations $a < b$ or $b < a$ or $a = b$ hold.

(iii) $a < b$ and $b < c$ imply $a < c$, for any a, b, c.

(iv) $a < b$ implies $a + c < b + c$, for any a, b, c.

(iv) $a < b$ and $0 < c$ imply $a \cdot c < b \cdot c$ and $c \cdot a < c \cdot b$, for any a, b, c.

Every ordered field contains the rational numbers as a subfield: $\mathbf{Q} \subseteq \mathbf{F}$. Hence every ordered field has the characteristic zero, and especially it is infinite. On each line, the points of a plane over an *ordered* field have a natural order (betweenness) relation. It turns out that Hilbert's relevant axioms of order hold, too.

Definition III.6.2. A point $B = (b_1, b_2)$ lies between a point $A = (a_1, a_2)$ and a point $C = (c_1, c_2)$ if and only if

$$b_i = \lambda a_i + (1 - \lambda) c_i \quad \text{for } i = 1, 2 \text{ with } 0 < \lambda < 1. \tag{III.6.1}$$

Proposition III.6.1. *In a Cartesian plane over an* ordered *field, Hilbert's axioms of order* (II.1)(II.2)(II.3)(II.4) *hold.*

Proposition III.6.2. *Conversely, assume that the axioms of order* (II.1) *through* (II.4) *hold in a Cartesian plane constructed over some field. Then one can introduce into this field an order compatible with the order of the plane.*

Remark. There exists an affine incidence plane for which the Euclidean parallel property (IV*), the axioms of order (II.1) through (II.4) all hold, but the Theorems of Desargues and Theorem of Pappus are both violated. In 1902, F.R. Moulton [4] has described such an example in his article *"A simple non-desarguesian plane geometry"*. Because of Theorem III.6.1 this is an example of an ordered affine incidence plane that is not a Cartesian plane over a field.

We are now ready to introduce congruence. From a Pythagorean field, we reconstruct a Pythagorean plane as the analytic geometry over the field. This construction works for an arbitrary Pythagorean field, and thus we continue our road from algebra to geometry. The introduction of Cartesian coordinates turns out to be a perfect way to produce many different models for Pythagorean planes. Recall the definitions:

Definition (II.3.2 Pythagorean plane). A *Pythagorean plane* is a Hilbert plane for which the Euclidean parallel postulate holds.

Definition (III.5.1 Pythagorean field). A *Pythagorean field* is an ordered field with the property

(Pyth) If $a, b \in \mathbf{F}$, then $\sqrt{a^2 + b^2} \in \mathbf{F}$.

This section gives the detailed direct proof of the following result:

Theorem III.6.2. *The Cartesian plane over a Pythagorean field is a Pythagorean plane.*

The converse result have already been obtained in the section on Hilbert's arithmetic of segments. By Theorem III.5.2, any Pythagorean plane is isomorphic to the Cartesian plane over its field of segment lengths, which is an Pythagorean field. Taking both results together, we state the important result:

Main Theorem 2 (Fundamental Theorem about Pythagorean Geometry). *The Cartesian plane* $\mathbf{F}^2$ *over a Pythagorean field* $\mathbf{F}$ *is a Pythagorean plane.*

Conversely, any Pythagorean plane is isomorphic to the Cartesian plane over its field of segment arithmetic, which is a Pythagorean field.

Hence there is a bijection between the Pythagorean planes and the Pythagorean fields.

The proof of the first part uses the fact that enough rigid motions exist in a Cartesian plane over a Pythagorean field. For the proof of the second part, one constructs the field with Hilbert's new arithmetic of segments *(Streckenrechnung)* as explained in the section "Arithmetic of Segments—Hilbert's Road from Geometry to Algebra".

Our next step towards more and more familiar geometry is to add some axiom justifying the classical use of ruler and compass.

Definition III.6.3 (Euclidean plane). A *Euclidean plane* is a Hilbert plane for which both the Euclidean parallel postulate and the circle-circle intersection property hold.

Definition III.6.4 (Euclidean field). A *Euclidean field* is an ordered field with the following property:

(3) If $a \in \mathbf{F}$ and $a \geq 0$, then $\sqrt{a} \in \mathbf{F}$.

Putting results together we get what I would be willing to name the fundamental theorem of Euclidean geometry:

Main Theorem 3 (Fundamental Theorem about Euclidean Geometry). *The Cartesian plane* $\mathbf{F}^2$ *over a Euclidean field* $\mathbf{F}$ *is a Euclidean plane.*

Conversely, any Euclidean plane is isomorphic to the Cartesian plane over its field of segment arithmetic, which is a Euclidean field.

Hence there is a bijection between the Euclidean planes and the Euclidean fields.

Definition III.6.5 (real closed field). A field is called *real closed* if every polynomial of odd order has at least one zero.

In a fourth step, we use Archimedes' axiom in order to justify measurements. Finally, we introduce a suitable axiom of completeness.

Definition III.6.6 (Archimedean plane). A *Archimedean plane* is a Hilbert plane for which the Archimedean Axiom hold.

Definition III.6.7 (Archimedean field). A field or skew field is called *Archimedean* if for any $a, c > 0$ there exists a natural number n such that $a < n \cdot c$.

Theorem III.6.3. *The Cartesian plane* $\mathbf{F}^2$ *is Archimedean if and only if the field* $\mathbf{F}$ *is Archimedean.*

Definition III.6.8 (Real Euclidean plane). A *real Euclidean plane* or simply *real plane* is a Hilbert plane for which both the Euclidean parallel postulate and the Dedekind axiom hold.

Putting results together we get what I would be willing to name the fundamental theorem of geometry:

Main Theorem 4 (Fundamental Theorem of Geometry). *The real Cartesian plane* $\mathbf{R}^2$ *is a complete, Archimedean Euclidean plane.*

Conversely, any complete, Archimedean Euclidean plane is isomorphic to the real Cartesian plane. The real numbers are obtained as its field of segment arithmetic.

We thus get a chain of models of strictly decreasing familiarity:

- The real Euclidean plane is a Euclidean plane.
- Any Euclidean plane is a Pythagorean plane.
- Any Pythagorean plane is a semi-Euclidean plane.

- Any semi-Euclidean plane is a Hilbert plane.
- Any Hilbert plane is an affine plane.
- Any affine plane is an incidence plane.

The converse does not hold in any of these steps.

III.6.2 About Archimedean Fields

Proposition III.6.3 (Theorem 59 of Hilbert). *Every Archimedean skew field is commutative.*

Proof. We get by an easy induction that

$$n \cdot a = a \cdot n$$

for any natural number n and any a of the skew field. Take two arbitrary elements a and b, for which we need to check the commutative law. It is easy to see that it is enough to consider—towards a contradiction—the case

$$a > 0\,,\ b > 0\,,\ ab - ba > 0 \tag{III.6.2}$$

From the existence of the inverse, we conclude that there exists an element $c > 0$ such that

$$(a + b + 1) \cdot c = ab - ba \tag{III.6.3}$$

There exists a number d such that

$$0 < d < 1\,,\ d < c$$

Recall that each nonempty set of natural numbers has a smallest one. Hence it is easy to see that there exist natural numbers $n, m \geq 0$ such that

$$m \cdot d < a \leq (m+1) \cdot d \quad \text{and} \quad n \cdot d < b \leq (n+1) \cdot d$$

Multiplication and subtraction yield

$$\begin{gathered} mn \cdot d^2 < ba < ab \leq mn \cdot d^2 + (m+n+1) \cdot d^2 \\ ab - ba < (m+n+1) \cdot d^2 \\ (m+n+1) \cdot d^2 < (a+b+1) \cdot d < (a+b+1)c \\ ab - ba < (a+b+1)c \end{gathered}$$

contradicting relation (V.12.1). Hence the assumption (III.6.2) cannot be true, and the commutative law is confirmed. □

III.6.3 Congruence in a Pythagorean Plane

For any two points $A = (a_1, a_2)$ and $B = (b_1, b_2)$, we introduce the distance

$$d(A, B) = \sqrt{(a_1 - b_1)^2 + (a_2 - b_2)^2}$$

and define the congruence of any two segments AB and $A'B'$ by postulating

$$AB \cong A'B' \Leftrightarrow d^2(A, B) = d^2(A', B')$$

From this definition one easily checks that congruence of segment is an equivalence relation and hence Hilbert's axiom (III.2) holds.

Problem III.6.1 (The rational plane is not Euclidean). *The rational Cartesian plane* $\mathbf{Q}^2$ *is an incidence plane, but not a Hilbert plane. Which are the two among Hilbert's axioms of incidence* (I.1)(I.2)(I.3), *and the axioms of order, congruence, parallelism, and continuity that do not hold in this case?*

We may assume to be known that $\sqrt{2}$ *is not rational. Explain with a simple counterexample involving* $\sqrt{2}$ *that one of the axioms of congruence does not hold.*

Answer. Only the axiom of congruence (III.1), and the completeness axiom (V.2) are not satisfied in the rational Cartesian plane $\mathbf{Q}^2$.

To see that the axiom of congruence (III.1) is violated, we take the unit square with vertices $(0,0), (1,0), (0,1), (1,1)$ and try to transfer its diagonal onto the positive x-axis, the ray $\overrightarrow{(0,0)(1,0)}$. This is impossible because a segment $(0,0)(\frac{p}{q}, 0)$ with length $\sqrt{2}$ does not exist.

The completeness axiom (V.2) is violated, since it is possible to adjoin additional points into the rational plane in such a way to obtain a Pythagorean plane.

Problem III.6.2. *Check that Hilbert's axiom* (III.1) *about the transfer of segments holds in a plane over a Pythagorean field.*

Proof. In axiom (III.1) are given arbitrary points A, B and A' and an arbitrary ray with vertex A'. We may suppose this ray is $a' = \overrightarrow{A'C}$ with $C \neq A'$. To check the claim made by the axiom, we need to prove existence of a point B' on this ray for which the segments $AB \cong A'B'$ are congruent.

With the coordinates

$$A = (a_1, a_2)\ ,\ B = (b_1, b_2)\ ,\ A' = (a'_1, a'_2)\ ,\ C = (c_1, c_2)$$

the equation of the ray is

$$x = a'_1 + \lambda(c_1 - a'_1) \quad \text{and} \quad y = a'_2 + \lambda(c_2 - a'_2) \quad \text{with} \quad \lambda \geq 0$$

To get the required congruence $AB \cong A'B'$, we need to satisfy the equation

$$(b_1 - a_1)^2 + (b_2 - a_2)^2 = (x - a_1')^2 + (y - a_2')^2$$

Plugging in the equation for the ray yields

$$(b_1 - a_1)^2 + (b_2 - a_2)^2 = \lambda^2 \left[(c_1 - a_1')^2 + (c_2 - a_2')^2\right]$$

It is possible to solve for the parameter $\lambda > 0$ since $C \neq A'$. We plug into the equation of the ray and get the coordinates of the required point $B' = (b_1', b_2')$ to be

$$b_1' = a_1' + \lambda(c_1 - a_1') \quad \text{and} \quad b_2' = a_2' + \lambda(c_2 - a_2') \quad \text{with } \lambda = \sqrt{\frac{(b_1 - a_1)^2 + (b_2 - a_2)^2}{(c_1 - a_1')^2 + (c_2 - a_2')^2}}$$

Hence we have proved existence of the point $B' = (b_1', b_2')$. □

Problem III.6.3 (adding two adjacent segments on a line). *Check that Hilbert's axiom* (III.3) *holds in a plane over a Pythagorean field.*

Proof. In axiom (III.3) are given two segments AB and BC lying on a line with B as the only point in common. In the same manner, two segments $A'B'$ and $B'C''$ are given on a line with only point B' in common. In other words, we assume that point $B = (b_1, b_2)$ lies between point $A = (a_1, a_2)$ and point $C = (c_1, c_2)$, and similarly for the primed items. Hence by definition III.6.2

$$\begin{aligned} b_i &= \lambda a_i + (1 - \lambda)c_i \quad \text{for } i = 1, 2 \text{ with } 0 < \lambda < 1. \\ b_i' &= \lambda' a_i' + (1 - \lambda')c_i' \quad \text{for } i = 1, 2 \text{ with } 0 < \lambda' < 1. \end{aligned}$$

The axiom (III.3) takes

$$AB \cong A'B' \quad \text{and} \quad BC \cong B'C$$

as assumption for the given segments. By definition, this means

$$\begin{aligned} (b_1 - a_1)^2 + (b_2 - a_2)^2 &= (b_1' - a_1')^2 + (b_2' - a_2')^2 \\ (b_1 - c_1)^2 + (b_2 - c_2)^2 &= (b_1' - c_1')^2 + (b_2' - c_2')^2 \end{aligned}$$

We plug in equation (III.6.3) and get

$$\begin{aligned} (1 - \lambda)^2[(a_1 - c_1)^2 + (a_2 - c_2)^2] &= (1 - \lambda')^2[(a_1' - c_1')^2 + (a_2' - c_2')^2] \\ \lambda^2[(a_1 - c_1)^2 + (a_2 - c_2)^2] &= \lambda'^2[(a_1' - c_1')^2 + (a_2' - c_2')^2] \end{aligned}$$

If either $A = C$ or $A' = C'$, both equalities hold, and nothing is to be proved. We can assume $A \neq C$ and $A' \neq C'$, and divide the two relations to get

$$\frac{(1-\lambda)^2}{\lambda^2} = \frac{(1-\lambda')^2}{\lambda'^2}$$

Since $0 < \lambda < 1$ and $0 < \lambda' < 1$, we obtain via the root

$$\frac{1-\lambda}{\lambda} = \frac{1-\lambda'}{\lambda'}$$
$$\lambda = \lambda'$$
$$(a_1 - c_1)^2 + (a_2 - c_2)^2 = (a_1' - c_1')^2 + (a_2' - c_2')^2$$
$$AC \cong A'C'$$

Hence we have proved

$$\text{if} \quad AB \cong A'B' \text{ and } \quad BC \cong B'C' \ , \quad \text{then} \quad AC \cong A'C'$$

as was claimed. □

III.6.4 Transfer of an Angle

Any ray h in the coordinate plane can be specified by its vertex (p_1, p_2) and a unit vector (h_1, h_2). Thus $h_1^2 + h_2^2 = 1$ is assumed. The points on the ray are given by the set

$$\{(p_1 + \lambda h_1, p_2 + \lambda h_2) \ : \ \lambda \geq 0\}$$

Recall the definition of an *orientated angle.*

Definition III.6.9 (Orientated angle). An *orientated angle* is an ordered pair (h, k) of two rays h and k with a common vertex. It is allowed that h and k are equal or opposite rays. To distinguish from a usual angle, I denote the orientated angle by $\curvearrowleft (h, k)$.

Definition III.6.10 (cosine and sin of an oriented angle). We define the cosine and sin of the orientated angle between two rays h and k to be

$$\cos \curvearrowleft (h, k) = h_1 k_1 + h_2 k_2 \ \text{ and } \ \sin \curvearrowleft (h, k) = h_1 k_2 - h_2 k_1$$

as elements of the field $\mathbf{F}$.

Remark. Take the case that the coordinate plane is constructed from a Pythagorean plane, as explained in the previous section. In that case, the above formulas for the measurement of the oriented angle may be checked using trigonometry. The calculations get easier using the imaginary unit i:

$$\cos \curvearrowleft (h,k) + i \sin \curvearrowleft (h,k) = h_1 k_1 + h_2 k_2 + i(h_1 k_2 - h_2 k_1) = (h_1 - i h_2)(k_1 + i k_2)$$

Let r be the ray from the origin along the positive x-axis. Let $\alpha = \curvearrowleft (r,h)$ and $\beta = \curvearrowleft (r,k)$ be the oriented angles between the positive x- axis and the ray h respectively k. In that case the above definition agrees with the usual definition of the trigonometric functions. Hence

$$\cos \curvearrowleft (h,k) + i \sin \curvearrowleft (h,k) = (\cos\alpha - i \sin\alpha)(\cos\beta + i \sin\beta) = \cos(\beta-\alpha) + i \sin(\beta-\alpha)$$

where we have used the addition theorems. Hence we get $\curvearrowleft (h,k) = \beta - \alpha$ for the measurement of the oriented angle, as it has to be.

In other words, the definition III.6.10 is in reality only a definition as long as it has not yet been confirmed that the coordinate plane over a Pythagorean field is a Pythagorean plane. After that confirmation, the above definition becomes legitimate. Indeed it turns out to be based on known trigonometric facts anyway. Of course, in case a contradiction would arise, and the above definition would be false and would have to be modified.

Definition III.6.11 (Congruence of orientated angles). Two oriented angles $\curvearrowleft (h,k)$ and $\curvearrowleft (h',k')$ are called *congruent* if and only if

$$\cos \curvearrowleft (h,k) = \cos \curvearrowleft (h',k') \quad \text{and} \quad \sin \curvearrowleft (h,k) = \sin \curvearrowleft (h',k')$$

As usual, an angle is defined to be an unordered pair of two rays with a common vertex that do not lie on a line. The notion of congruence of oriented angles is more restrictive than the congruence for the usual angles.

Definition III.6.12 (Congruence of angles). Two angles $\angle(h,k)$ and $\angle(h',k')$ are called *congruent* if and only if either

- the orientated angles $\curvearrowleft (h,k)$ and $\curvearrowleft (h',k')$ are congruent, or
- the orientated angles $\curvearrowleft (k,h)$ and $\curvearrowleft (h',k')$ are congruent.

These are all definition for a model of a Pythagorean plane.

Lemma III.6.1. *Given is any oriented angle $\curvearrowleft (h,k)$ and a ray h'. There exists a unique ray k' such that the angles $\curvearrowleft (h,k)$ and $\curvearrowleft (h',k')$ are congruent.*

Proof. The unit vectors (h_1, h_2), (k_1, k_2), and (h'_1, h'_2) are given. We have to find the unit vector such that the angles $\curvearrowleft (h, k)$ and $\curvearrowleft (h', k')$ are congruent. By definition that means we need to solve the linear system

$$\begin{aligned} h'_1 k'_1 + h'_2 k'_2 &= h_1 k_1 + h_2 k_2 \\ -h'_2 k'_1 + h'_1 k'_2 &= -h_2 k_1 + h_1 k_2 \end{aligned}$$

with the unknowns (k'_1, k'_2). This is a linear system and its determinant is ${h'_1}^2 + {h'_2}^2 = 1$. Hence there is a unique solution. We use Cramer's rule and get

$$\begin{aligned} k'_1 &= (h_1 k_1 + h_2 k_2) h'_1 - (-h_2 k_1 + h_1 k_2) h'_2 \\ k'_2 &= h'_1(-h_2 k_1 + h_1 k_2) - (-h'_2)(h_1 k_1 + h_2 k_2) \end{aligned}$$

as the unique solution. One needs still to check that this is a unit vector. This is just a lengthy calculations and turns out to be true.

$${k'_1}^2 + {k'_2}^2 = (h_1^2 + h_2^2)(k_1^2 + k_2^2) = 1$$

□

In every ordered incidence plane, it is possible to define and fix an orientation. The orientation is fixed, as soon as one has agreed which are the left and right half plane for one fixed ray. We establish an orientation in the coordinate plane, as usually done. The half-plane with $y > 0$ lies on the left side of the positively oriented x-axis.

Under the assumption that the rays h and k do not lie on a line, the angles $\curvearrowleft (h, k)$ and $\curvearrowleft (k, h)$ have different orientations. The orientations are different because of

$$\begin{aligned} \sin \curvearrowleft (h, k) &= -\sin \curvearrowleft (k, h) \neq 0 \\ \cos \curvearrowleft (h, k) &= \cos \curvearrowleft (k, h) \end{aligned}$$

Lemma III.6.2. *The unique angle transfer according to axiom* (III.4) *is valid in the reconstructed Pythagorean plane and indeed axiom* (III.4) *does hold.*

Indication of the solution. We need to transfer the two oriented angles $\curvearrowleft (h, k)$ and $\curvearrowleft (k, h)$ and get two solutions. Under the assumption that the rays h and k do not lie on a line, the angles $\curvearrowleft (h, k)$ and $\curvearrowleft (k, h)$ have different orientations. Hence the transferred angles have different orientations, too. We get transfers into the two opposite half-planes. Only the transfer of an angle into a given half-plane is unique. □

Lemma III.6.3. *Hilbert's axiom* (III.4) *about the transfer of angles holds in a plane over a Pythagorean field.*

Definition III.6.13 (Slope of a line). For a line AB through points $A = (a_1, a_2)$ and $B = (b_1, b_2)$, we define the *slope*

$$m = \frac{b_2 - a_2}{b_1 - a_1}$$

Opposite rays on the same line have the same slope. A vertical ray gets the slope $\infty \notin \mathbf{F}$, hence the slope of a line is an element in $\mathbf{F} \cup \{\infty\}$.

III.6.5 Enough Rigid Motions Exist in a Pythagorean Plane

Definition III.6.14 (Rigid motion). A *rigid motion* in a Pythagorean plane is a bijection ϕ from the set of points to itself and a bijection of the set of lines to itself which preserves incidence and order, and maps any segment AB or angle $\angle(h,k)$ to a congruent segment $\phi(A)\phi(B) \cong AB$ or angle $\angle(\phi(h), \phi(k)) \cong \angle(h,k)$.

Definition III.6.15 (Similarity). A *similarity* in a Pythagorean plane is a bijection ϕ from the set of points to itself and a bijection of the set of lines to itself which preserves incidence and order, and maps any angle $\angle(h,k)$ to a congruent angle $\angle(\phi(h), \phi(k)) \cong \angle(h,k)$, and any pair of congruent segments $AB \cong A'B'$ to a pair of congruent segments $\phi(A)\phi(B) \cong \phi(A')\phi(B')$.

Definition III.6.16 (Preserving versus reversing orientation). A rigid motion ϕ of any ordered affine plane preserves orientation if and only if any triangle $\triangle ABC$ and its image $\triangle\phi(A)\phi(B)\phi(C)$ always have the same orientation.

The rigid motion ϕ reverses orientation if and only if any triangle $\triangle ABC$ and its image $\triangle\phi(A)\phi(B)\phi(C)$ always have the opposite orientation.

Equivalently, we can state that a rigid motion preserves orientation if any ray$\overrightarrow{AB}$ and point C in the left half plane of this ray is mapped to a ray $\overrightarrow{\phi(A)\phi(B)}$ and point $\phi(C)$ in the left half plane of the image ray.

Theorem III.6.4 (There are enough rigid motions). *In a Pythagorean plane are given are two points P and Q, and rays r_P and r_Q with these vertices. There exist exactly <u>two</u> rigid motions ϕ and ψ, which map P to Q and r_P to r_Q. The first mapping ϕ preserves the orientation, and the second mapping ψ reverses the orientation.*

Lemma III.6.4. *For any point $A = (a_1, a_2)$, there exists a rigid translation which maps A to O, and which is orientation preserving.*

Lemma III.6.5. *The reflection P at the origin is a rigid motion which preserves the orientation.*

Lemma III.6.6. *The rotation by a right angle around the origin is a rigid motion which preserves the orientation.*

Problem III.6.4. *Check that the linear equations*

$$\begin{aligned} x' &= cx + sy \\ y' &= -sx + cy \end{aligned}$$

with $c^2 + s^2 = 1$ define an isometry in the Pythagorean plane.

Problem III.6.5. *Given the orientated angle $\angle EOA$ with $E = (1,0), O = (0,0), A = (x,y)$ and $y > 0$. Check that the linear equations*

$$\begin{aligned} x' &= cx + sy \\ y' &= -sx + cy \end{aligned}$$

with $c^2 + s^2 = 1, c > 0, s > 0$ map the orientated angle $\angle EOA$ to a congruent angle.

Answer. The sides $\overrightarrow{OE}$ and $\overrightarrow{OA}$ of the angle $\angle EOA$ have the slopes $0 = m_p$ and $y/x = m_q$. Hence

$$\tan \curvearrowleft (p,q) = \frac{m_q - m_p}{1 + m_p m_q} = \frac{y}{x}$$

Point $E = (1,0)$ is mapped to $E' = (c, -s)$. Hence the ray $\overrightarrow{OE'}$ has the slope

$$m_p' = -\frac{s}{c}$$

The other image ray $\overrightarrow{OA'}$ has the slope

$$m_q' = \frac{y'}{x'} = \frac{-sx + cy}{cx + sy}$$

Hence

$$\tan \curvearrowleft (p', q') = \frac{m_q' - m_p'}{1 + m_p' m_q'} = \frac{(-sx + cy)c + s(cx + sy)}{c(cx + sy) - s(-sx + cy)} = \frac{y}{x}$$

We can also check that both the angle $\angle EOA$ and its image $\angle E'OA'$ have the same orientation. This is easy enough for the special case $A = (0,1)$, from which the general case follows. Hence the angle $\angle EOA$ and its image $\angle E'OA'$ are congruent.

Lemma III.6.7. *The linear equations*

$$x' = cx + sy$$
$$y' = -sx + cy$$

with $c^2 + s^2 = 1$ *define a rigid rotation in the Pythagorean plane.*

Lemma III.6.8. *The reflection* I_x *across the* x*-axis is a rigid motion which reverses the orientation.*

Lemma III.6.9. *Given is a Pythagorean plane. For any ray* $\overrightarrow{OA}$ *pointing into the right half plane, there exists a rigid rotation which maps this ray into the positive* x*-axis and preserves orientation.*

Proof. Let $A = (a_1, a_2)$ be the coordinates of A, point $E = (1, 0)$ and

$$m = \tan \angle EOA = \frac{a_2}{a_1}$$

be the slope of ray $\overrightarrow{OA}$. In a Pythagorean field, we can get the cosine and sinus of the rotation angle $\angle EOA$ as

$$c = \frac{1}{\sqrt{1+m^2}} = \frac{a_1}{a_1^2 + a_2^2} \quad \text{and} \quad s = \frac{m}{\sqrt{1+m^2}} = \frac{a_2}{a_1^2 + a_2^2}$$

These quantities exist: Indeed, the square root $\sqrt{1+m^2}$ exists in a Pythagorean field. Moreover, $1 + m^2 \geq 1$ cannot be zero, hence $\sqrt{1+m^2} \neq 0$ and the division by this quantity is possible. The linear equations

$$x' = cx + sy$$
$$y' = -sx + cy$$

can be checked to be the rigid rotation mapping the ray $\overrightarrow{OA}$ to the positive x-axis $\overrightarrow{OE}$. □

Lemma III.6.10. *Any composition of rigid motions, and the inverse of a rigid motion is again a rigid motion. In other words, the rigid motion are a group. The orientation preserving motions are a subgroup.*

Proof of the Theorem about enough rigid motions. By composition of a translation and a rotation, a ray towards right can be mapped to the positive x-axis. For an arbitrary ray, the same result can be achieved by use of a point reflection or a rotation by a right angle as third factor. We thus obtain an orientation preserving rigid motion with the required property. Composition with a reflection across the x-axis yields an orientation reversing rigid motion with the required property. □

Corollary 12. *Every rigid motion can be realized by a composition a translation, a rotation about the origin, and possibly a reflection across the x-axis.*

Proposition III.6.4. *In a Pythagorean coordinate plane are given two triangles $\triangle ABC$ and $\triangle A'B'C'$. If the congruences*

$$AB \cong A'B' \, , \; AC \cong A'C' \, , \; \angle BAC \cong \angle B'A'C'$$

hold, then the two triangles are congruent. Especially Hilbert's SAS-axiom (III.5) *holds in a plane over a Pythagorean field.*

Proof. I restrict myself to the case that angle $\angle BAC$ is acute. There exists a rigid motion ϕ mapping A to O, the ray $\overrightarrow{AB}$ to the positive x-axis $\overrightarrow{OE}$, and vertex C to a point in the upper half plane. Hence

$$\phi(A) = O = (0,0) \, , \, \phi(B) = (0,c) \quad \text{and} \quad \phi(C) = (x,y) \quad \text{with} \quad y > 0$$

The rigidity of the mapping ϕ implies

$$\begin{aligned} c = d(A,B) &= \sqrt{(a_1-b_1)^2 + (a_2-b_2)^2} \, , \\ x^2 + y^2 = d^2(A,C) &= (a_1-c_1)^2 + (a_2-c_2)^2 \quad \text{and} \\ \frac{y}{x} &= \tan \angle BAC \end{aligned}$$

All corresponding claims holds for a mapping ϕ' related to the second triangle $\triangle A'B'C'$. Comparing the relations obtained for both triangles, and using the congruences assumed, we get

$$\begin{gathered} c = d(A,B) = d(A',B') = c' \\ x^2 + y^2 = d^2(A,C) = d^2(A',C') = x'^2 + y'^2 \\ \frac{y}{x} = \tan \angle BAC = \tan \angle B'A'C' = \frac{y'}{x'} \\ y > 0 \;\text{ and } y' > 0 \end{gathered}$$

From the assumption that angle $\angle BAC \cong \angle B'A'C'$ is acute, we get $x > 0$ and $y > 0$ and similarly $x' > 0$ and $y' > 0$. Hence we conclude $x = x', y = y'$ and $c = c'$. Hence the two "moved triangles" $\triangle\phi(A)\phi(B)\phi(C)$ with the vertices

$$\phi(A) = (0,0) \, , \, \phi(B) = (0,c) \quad \text{and} \quad \phi(C) = (x,y)$$

and the corresponding primed one are <u>equal</u>. Hence they are congruent, too! Since a rigid motions maps a triangle to a congruent one, we conclude that the original triangles $\triangle ABC \cong \triangle A'B'C'$ are congruent, too, as to be shown.

In the case of an obtuse angle $\angle BAC \cong \angle B'A'C'$, we get $x > 0$ and $y < 0$ and similarly $x' > 0$ and $y' < 0$. Thus $x = x', y = y'$ follows similarly to above. The proof is even easier for the case that angle $\angle BAC \cong \angle B'A'C'$ is right. □

End of the proof of Theorem III.6.2. We finish the proof that a Cartesian plane over a Pythagorean field is a Pythagorean plane.

- By Theorem II.4.2, a Cartesian plane over a field is an affine plane. This means that Hilbert's axioms (I.1)(I.2)(I.3a)(I.3b) and the Euclidean parallel property hold.
- By Proposition III.6.1, in a Cartesian plane over an *ordered* field, Hilbert's axioms of order (II.1)(II.2)(II.3)(II.4) hold.
- In the solution of Problem III.6.2, we have checked that Hilbert's axiom (III.1) about the transfer of segments holds in a plane over a Pythagorean field.
- From the definition of congruence of segments, one sees that congruence is an equivalence relation, and hence Hilbert's axiom (III.2) holds.
- In the solution of Problem III.6.3, we have checked that Hilbert's axiom (III.3) about segment addition holds in a plane over a Pythagorean field.
- By Lemma III.6.3, Hilbert's axiom (III.4) about the transfer of angles holds in a plane over a Pythagorean field.
- By Proposition III.6.4, Hilbert's SAS-axiom (III.5) holds in a plane over a Pythagorean field.

Thus all five axioms of congruence (III.1) through (III.5) are satisfied in a Cartesian plane over a Pythagorean field. □

III.6.6 Euclidean Fields and Intersection Properties of Circles

Theorem III.6.5. *The following three properties are equivalent in any Pythagorean coordinate plane* $\mathbf{F}^2$*:*

(1) *The field* $\mathbf{F}$ *is Euclidean.*

(2) *The circle-line intersection property holds in the plane* $\mathbf{F}^2$.

(3) *The circle-circle intersection property holds in the plane* $\mathbf{F}^2$.

Lemma III.6.11. *A Euclidean field is Pythagorean.*

Lemma III.6.12 ((1) ⇒ (2)). *The circle-line intersection property holds in the plane over a Euclidean field.*

Proof of (1) ⇒ (2). Given is a circle and a line which contains an interior point of the circle. Let P be the foot point of the perpendicular p dropped from the center of the circle onto the line. Since a Euclidean field is Pythagorean, there exist enough rigid motions in the plane $\mathbf{F}^2$. Hence there exists a rigid motion mapping the ray $\overrightarrow{OP}$ onto the positive x-axis. In the special case $O = P$ any one of the two rays with vertex perpendicular to the given line will be mapped onto the positive x-axis.

Thus we have put the circle with radius $c > 0$ and the line into a Cartesian coordinate system, with coordinates $C = (0, 0)$ for the center and $P = (p, 0)$ for the foot point.

Question. What are the equations for the circle and the line.

Answer.

$$\begin{aligned} x^2 + y^2 &= c^2 \\ x &= p \end{aligned} \tag{III.6.4}$$

Since the line contains a point (p, q) inside the circle, we know that $p^2 \leq p^2 + q^2 < c^2$ and hence $c^2 - p^2 > 0$.

Question. What are the coordinates (x, y) of the intersection points of the circle and the line.

Answer. Plugging $x = p$ into the equation of the circle yields $y^2 = c^2 - p^2$. We have seen above that $c^2 - p^2 > 0$ follows since the line contains a point inside the circle. The field is assumed to be Euclidean, hence we can take the square root of a positive quantity and get the two solutions $y = \pm\sqrt{c^2 - p^2}$. We have obtained two intersection points with the coordinates $(p, \sqrt{c^2 - p^2})$ and $(p, -\sqrt{c^2 - p^2})$.

We have checked that the circle-line intersection property holds in the coordinate plane over a Euclidean field. □

Lemma III.6.13 ((2) ⇒ (1)). *If the circle-line intersection property holds for the Pythagorean plane* $\mathbf{F}^2$, *then the field* $\mathbf{F}$ *is Euclidean.*

Proof of (2) $\Rightarrow$ (1). Given is $f > 0$, a positive element of the field. We construct in the plane $\mathbf{F}^2$ the circle and the line with the equations III.6.4 with

$$c = \frac{1+f}{2} \quad \text{and} \quad p = \frac{f-1}{2}$$

The assumption $f > 0$ implies $-c < p < c$. Hence the point $(p, 0)$ lies inside the circle. By assumption, the circle-line intersection property holds. Hence circle and line intersect in some point (p, y). From the equation of the circle and the assumption we get

$$y^2 = c^2 - p^2 = f$$

Hence $y = \sqrt{f}$ is a square root of f, the existence of which was to show. □

Lemma III.6.14 ((1) $\Rightarrow$ (3)). *The circle-circle intersection property holds in the plane over a Euclidean field.*

Proof of (1) $\Rightarrow$ (3). The two circles $\mathcal{C}$ with center C and radius c, and $\mathcal{D}$ with center D and radius d are given. We assume that one of the circles has points both inside and outside the other one.

Question. Which inequalities hold between the two radius $c, d > 0$ and the distance $a = |CD|$ between the two centers. Which case occurs in Proposition V.12.22.

Answer. If $a \geq c + d$, the interiors of the two circles are disjoint, which is case **(i)** of Proposition V.12.22. If $a \leq |c - d|$, the interior of the smaller circle lies entirely in the interior of the larger one. These are the cases **(ii)** and **(iii)** of Proposition V.12.22. These cases are excluded by the assumption that one circle has points both inside and outside the other one. Hence

$$|c - d| < a < c + d$$

These inequalities correspond in Proposition V.12.22 to only remaining possibility **(iv)**.

Since a Euclidean field is Pythagorean, there exist enough rigid motions in the plane $\mathbf{F}^2$. Hence there exists a rigid motion mapping the ray $\overrightarrow{CD}$ between the two centers onto the positive x-axis.

Question. What are the coordinates of the two centers of the circles.

Answer. We have put the two circle in a Cartesian coordinate system, with coordinates of the centers $C = (0, 0)$ and $D = (a, 0)$, where $a = |CD|$ is the distance of the two centers.

Question. What are the equations for the two circles.

Answer.

$$x^2 + y^2 = c^2$$
$$(x-a)^2 + y^2 = d^2$$

Question. What can one conclude from subtracting the two equations about the possible intersection of the two circles?

Answer. Substraction of the two equations yields a linear equation with one unknown x. One can even solve for x.

$$x = \frac{a^2 + c^2 - d^2}{2a} \tag{III.6.5}$$

This is the equation of a line. Hence the intersection points of the two circles lie on this line l.

Conversely, the intersection points of the line l with either circle lie on the other circle, too. Hence it is now enough to look for the intersection points of circle $\mathcal{C}$ and line l. We may assume that $c \geq d$, and use the abbreviation

$$p = \frac{a^2 + c^2 - d^2}{2a}$$

The point P with coordinates $(p, 0)$ lies on the line l.

Question. Why do the circle $\mathcal{C}$ and the line l intersect?

Answer. The point P lies inside the circle $\mathcal{C}$ since

$$-c < p < c$$

The right inequality can be checked as follows:

$$\begin{aligned}
\frac{a^2 + c^2 - d^2}{2a} &< c \\
a^2 + c^2 - d^2 &< 2ac \\
a^2 + c^2 - 2ac &< d^2 \\
(a-c)^2 &< d^2 \\
-d < a - c &< d \\
c - d < a &< c + d
\end{aligned}$$

The last of these inequalities is true by assumption, and they are all equivalent. Checking that $c + p > 0$ is a quite similar calculation left to the reader. Since the line l has the point P inside circle $\mathcal{C}$, the circle-line intersection property guarantees that they intersect.

Question. What are the coordinates (x, y) of the intersection points of the circle and the line.

Answer. Plugging $x = p$ into the equation of the first circle yields $y^2 = c^2 - p^2$. We have seen above that $c^2 - p^2 = (c+p)(c-p) > 0$. The field is assumed to be Euclidean, hence we can take the square root of a positive quantity y^2 and get the two intersection points with the coordinates $(p, \sqrt{c^2 - p^2})$ and $(p, -\sqrt{c^2 - p^2})$.

Thus we have that the circle-circle intersection property holds for a Euclidean plane. ☐

End of the proof of Theorem III.6.5. We know from Proposition V.12.20 that the circle-circle intersection property implies the line-circle intersection property. This holds even more generally in every Hilbert plane. Thus we know that (3) ⇒ (2) for the items of Theorem III.6.5. We have proved that (2) ⇒ (1)— the circle-line intersection property implies that field is Euclidean— in Lemma III.6.13 Moreover, Lemma III.6.14 states that (1) ⇒ (3)—The circle-circle intersection property holds in the plane over a Euclidean field. Thus all three statements (1)(2)(3) of Theorem III.6.5 are equivalent, as to be shown. ☐

Problem III.6.6 ((**3**) ⇒ (**1**))**.** *Give a direct proof of the statement that a Pythagorean plane* $\mathbf{F}^2$ *for which the circle-circle intersection property holds has a Euclidean field* $\mathbf{F}$.

Proof of (3) ⇒ (1). Given is $f > 0$, a positive element of the field. We construct in the plane $\mathbf{F}^2$ two circles with the same radius c and the distance a between there centers. We choose

$$c = \frac{1+f}{2} \quad \text{and} \quad a = |f - 1|$$

From the equations of the two circles

$$\begin{aligned} x^2 + y^2 &= c^2 \\ (x-a)^2 + y^2 &= c^2 \end{aligned}$$

we get the coordinates (x, y) of an intersection point to be $x = \frac{a}{2}$ and

$$y = \sqrt{y^2} = \sqrt{\left(\frac{1+f}{2}\right)^2 - \left(\frac{1-f}{2}\right)^2} = \sqrt{f}$$

Since the intersection point is assumed to exist, we see that the square root $\sqrt{f}$ exists for any $f > 0$, as to be shown. ☐

From the results above, we get a proof of Proposition III.6.5 as promised in the section on the neutral geometry of circles.

Proposition III.6.5. *In any Hilbert plane, the line-circle intersection property together with the parallel axiom imply the circle-circle intersection property.*

Reason. Assuming the parallel axiom, we may invoke theorem III.5.2: any Pythagorean plane is isomorphic to the Cartesian plane over its field of segment lengths, which is an Pythagorean field $\mathbf{F}$. Because the circle-line intersection property is assumed, we know from Lemma III.6.13 above, that the field $\mathbf{F}$ is even Euclidean. Hence by Lemma III.6.14, the circle-circle intersection holds, too. □

Corollary 13. *If the distance $a = |CD|$ between the centers of two circles $\mathcal{C}$ with center C and radius c, and $\mathcal{D}$ with center D and radius d satisfy the inequalities*

$$|c - d| < a < c + d$$

they intersect in two points.

Corollary 14 (Heron's formula). *In Euclidean geometry, the area of a triangle with the sides a, b and c is*

$$\text{area}(\triangle ABC) = \frac{1}{4}\sqrt{(a+b+c)(a+b-c)(a-b+c)(b+c-a)}$$

Proof. I replace b by d as always above and consider the triangle with the vertices C, D and an intersection point $X = (p, y)$ of the circles $\mathcal{C}$ and $\mathcal{D}$. We get the Euclidean area

$$\text{area}(\triangle CDX) = \frac{ay}{2} = \frac{a}{2}\sqrt{c^2 - p^2}$$

The definition of p implies

$$\begin{aligned} c - p &= \frac{2ac - a^2 - c^2 + d^2}{2a} = \frac{d^2 - (a-c)^2}{2a} = \frac{(d-a+c)(d+a-c)}{2a} \\ c + p &= \frac{2ac + a^2 + c^2 - d^2}{2a} = \frac{(a+c)^2 - d^2}{2a} = \frac{(a+c+d)(a+c-d)}{2a} \end{aligned} \tag{III.6.6}$$

The coordinates of the intersection points are $(x, y) = (p, \pm\sqrt{c^2 - p^2})$ with

$$p = \frac{a^2 + c^2 - d^2}{2a}$$

$$c^2 - p^2 = \frac{(a+c+d)(a+c-d)(a-c+d)(c-a+d)}{4a^2}$$

] Hence the area of the triangle with sides a, c and d is
(III.6.7)

$$\text{area}(\triangle CDX) = \frac{a}{2}\sqrt{c^2 - p^2} = \frac{1}{4}\sqrt{(a+c+d)(a+c-d)(a-c+d)(c-a+d)}$$

□

Problem III.6.7. *Confirm that the line point P and the line l are constructible with Hilbert tools and unique parallels, even if no circle intersection properties are assumed.*

Proof. From the expression III.6.6 for $c - p$ and $c + p$, one gets the proportions

$$\frac{c-p}{a+c+d} = \frac{c+d-a}{2a} \tag{III.6.8}$$

$$\frac{c+p}{a+c-d} = \frac{a+c+d}{2a} \tag{III.6.9}$$

From these proportions, construction of either $c-p$ or $c+p$ can be implemented with Hilbert tools, using parallels and similar triangles.

Since the line l has the point P inside circle $\mathcal{C}$, the circle-line intersection property guarantees that they intersect. By the remarks above, there intersection points are the same as the intersection points of the two circles $\mathcal{C}$ and $\mathcal{D}$. □

Here is a "fairly clean" solution for the construction of point P. Let AB and FE be the diameters of the two given circles $\mathcal{C}$ and $\mathcal{D}$ lying on the line CD.

Question. We assume the order $A*F*B*E$. Calculate the distances $|AF|, |FB|, |BE|$ and $|AE|$.

Answer.

$$\begin{aligned} |AF| &= a + c - d \\ |FB| &= c + d - a \\ |BE| &= a - c + d \\ |AE| &= a + c + d \end{aligned}$$

We construct a rectangle with one side AB, and the perpendicular sides congruent to $2a$. The coordinates of its vertices are

$$A = (-c, 0), B = (c, 0), H = (c, 2a), G = (-c, 2a)$$

We draw a line g with slope -1 through point $F = (a-d, 0)$. It intersects the vertical sides of the rectangle, or their extensions, in the points

$$I = (-c, a+c-d) \quad \text{and} \quad J = (c, -(c+d-a))$$

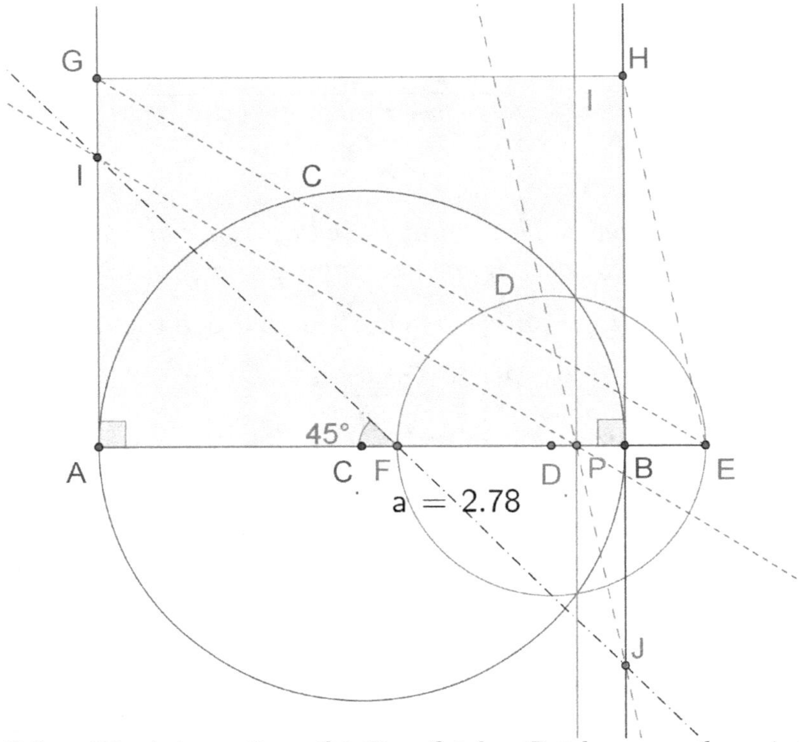

Figure III.6.1: The intersections $\mathcal{C} \cap \mathcal{D} = \mathcal{C} \cap l = \mathcal{D} \cap l$ are equal—point P and line l can be constructed with Hilbert tools.

The parallel to line GE through point I intersects the x-axis in point $P = (p, 0)$, because of the proportion (III.6.9) and the similar triangles $\triangle API \sim \triangle AEG$. With a similar reasoning, we get that the parallel to line HE through point J intersects the x-axis in the same point $P = (p, 0)$, too.

Part IV

Real Progress in Euclidean Geometry

IV.7 Similar Triangles

IV.7.1 Basic Properties from Euclid

Definition IV.7.1 (Similar triangles). Two triangles $\triangle ABC$ and $\triangle A'B'C'$ for which the ratios of corresponding sides

$$\frac{a}{b} = \frac{a'}{b'}, \; \frac{b}{c} = \frac{b'}{c'}, \text{ and } \; \frac{c}{a} = \frac{c'}{a'}, \tag{IV.7.1}$$

are equal, are called *similar*.

Hilbert's—and already Legendre's—view about the theory of similarity is different from Euclid's: In Hilbert's foundations, the results about similar triangles follow from the segment arithmetic. The segment arithmetic, in turn, is justified by Pappus theorem. Following Hilbert's approach, the properties of similar triangles shall be justified on the basis of the segment arithmetic developed in the previous section.

Proposition IV.7.1. *For any two similar triangles, there exists a* scaling factor f *by which one needs to multiply the length of a side of the triangle $\triangle ABC$ to obtain the length of the corresponding side of the equiangular triangle $\triangle A'B'C'$.*

Proof. By ordinary calculation with fractions,

$$\frac{a'}{b'} = \frac{a}{b} \quad \text{implies} \quad \frac{a'}{a} = \frac{b'}{b}$$

Similarly we get $\frac{b'}{b} = \frac{c'}{c}$. This common ratio is the scaling factor, which I call f and obtain

$$a' = fa\,,\; b' = fb\,, \quad \text{and} \quad c' = fc$$

□

Definition IV.7.2 (Equiangular triangles). Two triangles $\triangle ABC$ and $\triangle A'B'C'$ for which the angles across the corresponding sides are congruent are called *equiangular.*

Theorem IV.7.1. *Two triangles are equiangular if and only if they are similar.*

To prove this basic theorem, we begin with right triangles.

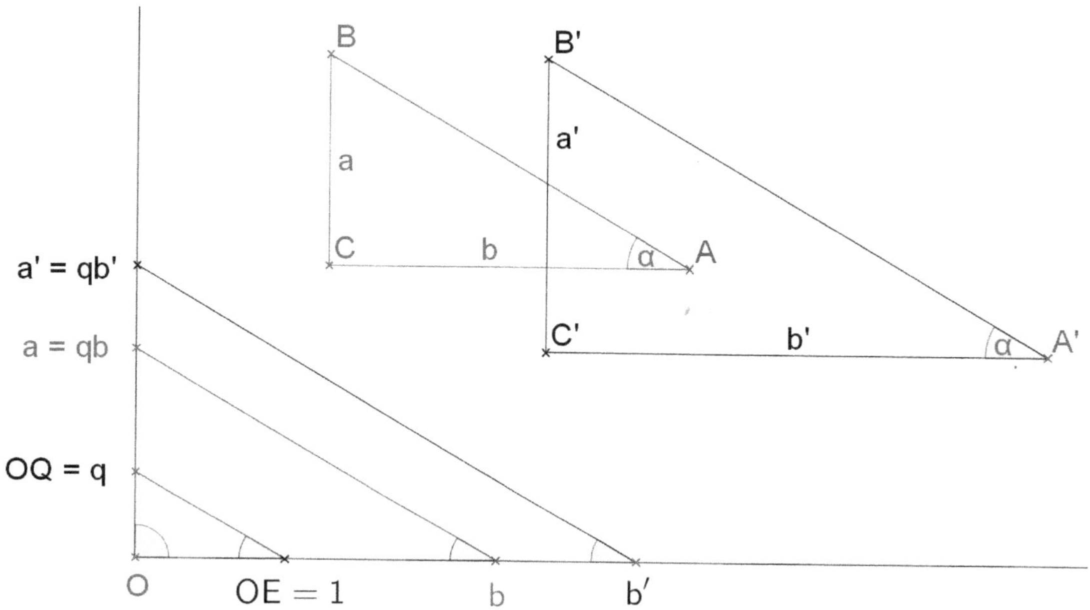

Figure IV.7.1: The legs of equiangular right triangles are proportional.

The legs of two equiangular right triangles have the same ratio. Given are two equiangular right triangles. We use SAS congruence to get triangles $\triangle OAB$ and $\triangle OA'B'$ congruent to the original ones, which have a common vertex O at the right angle, and the legs $|OB| = a, |OB'| = a'$ and $|OA| = b, |OA'| = b'$ lying on two perpendicular rays from this vertex. For convenience, we take these two rays horizontally to the right and upwards.

We draw the unit segment $|OE| = 1$ on the horizontal ray and the parallel to the hypothenuse AB through point E. We obtain the intersection point Q on the vertical ray. By segment arithmetic, the segment $|OQ| = q$ on the vertical ray satisfies

$$qb = a$$

Because the two hypothenuses AB and $A'B'$ have congruent angles α with the horizontal axis, the lines EQ and $A'B'$ are parallel, too. Hence with the same value of q, segment arithmetic now shows that

$$qb' = a'$$

and a simple calculation shows that

$$\frac{a}{b} = q = \frac{a'}{b'}$$

□

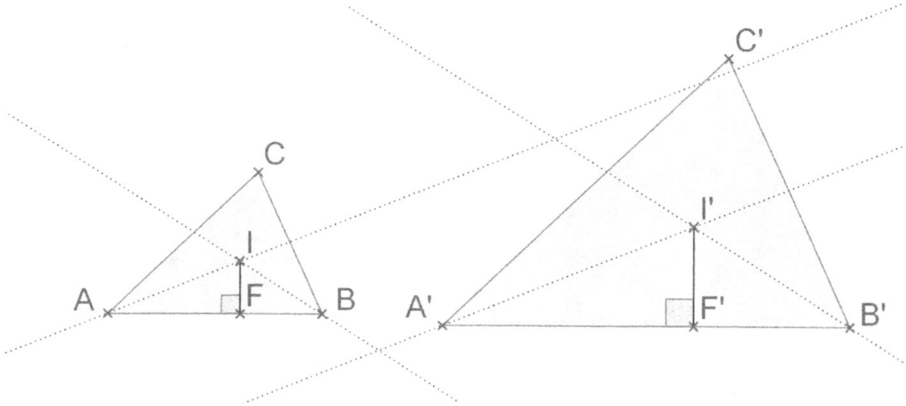

Figure IV.7.2: Equiangular triangles are similar.

Since we have shown above that the ratio $\frac{a}{b}$ depends only on the angles of the right triangle, can can define a trigonometric function:

Definition IV.7.3 (Tangent function for acute angles). The ratio $\frac{a}{b}$ of the leg a across to the leg b adjacent to the angle α of a right triangle is the *tangent function* of this angle:

$$\tan \alpha = \frac{a}{b}$$

Proposition IV.7.2 (Euclid VI.4). *The corresponding sides of equiangular triangles have equal ratios. Equiangular triangles are* similar. *In formulas:*

$$\frac{a}{b} = \frac{a'}{b'}, \quad \frac{b}{c} = \frac{b'}{c'} \quad \textit{and} \quad \frac{c}{a} = \frac{c'}{a'}$$

The proof uses some facts from neutral geometry about the in-circle of a triangle.

Theorem IV.7.2 (The in-circle). *All three interior angular bisectors of a triangle intersect in one point. This point has congruent distances from all three sides of the triangle, and hence it is the center of the in-circle.*

Problem IV.7.1 (The in-circle). *We want to construct the in-circle of a triangle. Explain the steps of the reasoning and provide drawings.*

(a) *Explain why any two interior angle bisectors of a triangle intersect.*

(b) *Explain by SAA triangle congruence why a point on the interior angular bisector has congruent distances from both sides of the angle*

(c) *Explain why this point has congruent distances from all three sides of the triangle.*

(d) *Explain why it is the center of the in-circle.*

(e) *Explain why all three interior angular bisectors of a triangle intersect in one point.*

(f) *Explain how one can draw the in-circle of a triangle.*

Answer. Here is some rather elaborate recollection:

(a) Why any two interior angle bisectors of a triangle intersect: By the Crossbar Theorem, an interior angular bisector intersects the opposite side. Let W_a be the intersection point of the angular bisector of angle $\angle BAC$ with side BC.

Once more, by the Crossbar Theorem, the angular bisector w_b of angle $\angle ABC$ intersects segment AW_a, since points A and W_a lie on the two sides of this angle. The bisector w_b intersects segment AW_a of bisector w_a, say in point I.

(b) is left to the reader.

(c) Point I has congruent distances from all three sides: Since point I lies one the bisector of angle $\angle BAC$, it has congruent distances from the sides AC and AB. Since point I lies one the bisector of angle $\angle ABC$, it has congruent distances from the sides BC and BA. By transitivity, it has congruent distances from all three sides.

(d) Point I is the center of the in-circle: Let X be the footpoint of the perpendicular dropped from point I onto the side BC. This side is tangent to the circle around I through X, since the tangent of a circle is perpendicular to the radius IX at the touching point. Indeed, we have already obtained a circle to which <u>all three sides</u> of the triangle are tangent. Let Y, Z be the footpoints of the of the perpendiculars dropped from point I onto the other two sides CA and AB. Since $IX \cong IY \cong IZ$, the same circle goes through Y and Z. Again, the sides CA and AB are tangent to this circle since they are perpendicular to IY and IZ, respectively.

(e) **All three interior angular bisectors of a triangle intersect in one point:** Since point I has congruent distances from the sides AC and BC, the point lies on either one of the bisectors of the angles at C. This is shown with the hypothenuse-leg congruence. Since point I lies in the interior of the angle $\angle BCA$, it lies on the interior angular bisector of this angle.

(f) **How to draw the in-circle of a triangle:** As seen above, one needs to construct the angular bisectors of any two of the three angles of the triangle. One drops the perpendicular from their intersection point I to any side of the triangle. Finally, one draws the circle around I through the foot point.

Proof of proposition IV.7.2. Let the two equiangular triangles $\triangle ABC$ and $\triangle A'B'C'$ be given. By Theorem V.12.4 from the section about neutral triangle geometry, every triangle has an in-circle. Its center I is the intersection point of the three interior angular bisectors. We drop the perpendicular from the center I onto the triangle side AB and let F be the foot point. We proceed similarly for the equiangular triangle $\triangle A'B'C'$. Thus we get two pairs of equiangular right triangles: $\triangle AIF$ is equiangular with $\triangle A'I'F'$, and $\triangle BIF$ is equiangular with $\triangle B'I'F'$. We get the proportions for the corresponding legs, which we have to add:

$$\frac{|AF|}{|FI|} = \frac{|A'F'|}{|F'I'|} \quad \text{and} \quad \frac{|FB|}{|FI|} = \frac{|F'B'|}{|F'I'|},$$
$$\text{and hence} \quad \frac{|AB|}{|FI|} = \frac{|A'B'|}{|F'I'|}$$

Now $|FI| = \rho$ and $|F'I'| = \rho'$ are the radii of the in-circles. For the sides $|AB| = c$ and $|A'B'| = c'$, as well as the other corresponding sides we obtain

$$\frac{c}{\rho} = \frac{c'}{\rho'}, \quad \frac{a}{\rho} = \frac{a'}{\rho'} \quad \text{and} \quad \frac{b}{\rho} = \frac{b'}{\rho'},$$
$$\text{Hence} \quad \frac{a}{b} = \frac{a'}{b'}, \quad \text{and similarly} \quad \frac{b}{c} = \frac{b'}{c'} \quad \text{and} \quad \frac{c}{a} = \frac{c'}{a'}$$

□

Remark. By the sin Theorem IV.9.1 proved below, the ratio of any two sides of a triangle is the ratio of the sinus of the opposite sides:

$$\frac{a}{b} = \frac{\sin\alpha}{\sin\beta}, \quad \frac{b}{c} = \frac{\sin\beta}{\sin\gamma}, \quad \text{and} \quad \frac{c}{a} = \frac{\sin\gamma}{\sin\alpha},$$

Hence we see once more why these ratios depend only of the angles of the triangle.

Proposition IV.7.3. *For any two equiangular triangles, there exists a scaling factor f by which one needs to multiply the length of a side of the triangle $\triangle ABC$ to obtain the length of the corresponding side of the equiangular triangle $\triangle A'B'C'$:*

$$a' = fa\,,\ b' = fb\,, \quad \text{and} \quad c' = fc$$

Proof. As explained above, the scaling factor is the common ratio $f := \frac{a'}{a} = \frac{b'}{b} = \frac{c'}{c}$. □

Proposition IV.7.4 (Euclid VI.5). *Similar triangles are equiangular.*

Proof. Given are two triangles $\triangle ABC$ and $\triangle A'B'C'$ the lengths of the sides of which satisfy

$$\frac{a}{b} = \frac{a'}{b'}\,,\ \frac{b}{c} = \frac{b'}{c'} \quad \text{and hence} \quad \frac{a}{c} = \frac{a'}{c'}$$

As shown earlier, in a Hilbert plane with unique parallels, there exist equiangular triangles of arbitrary sizes. See the section on the natural axiomatization, Wallis' postulate in volume II.

We transfer the angle $\angle BAC$ to vertex A' and angle $\angle ABC$ to vertex B', each of both with one side along line $A'B'$. We obtain two new rays, both in the same half plane as point C'. The two new rays intersect at a point X. [9] We obtain the equiangular triangles $\triangle ABC$ and $\triangle A'B'X$, with the (possibly new) vertex X,— and both $\triangle A'B'X$ and $\triangle A'B'C'$ lying in the same half plane of line $A'B'$.

By Euclid VI.4 (stated in proposition IV.7.2), the equiangular triangles $\triangle ABC$ and $\triangle A'B'X$ are similar. Let $|B'X| = a''$ and $|A'X| = b''$. We have obtained the proportions

$$\begin{aligned} \frac{b}{c} &= \frac{b''}{c'} \quad \text{and} \quad \frac{a}{c} = \frac{a''}{c'} \\ \text{Hence} \quad \frac{b'}{c'} &= \frac{b''}{c'} \quad \text{and} \quad \frac{a'}{c'} = \frac{a''}{c'} \\ \text{Hence} \quad b' &= b'' \quad \text{and} \quad a' = a'' \end{aligned}$$

We see that triangles $\triangle A'B'C'$ and $\triangle A'B'X$ have three pairs of congruent sides. Hence the two triangles are congruent by SSS-congruence Theorem V.12.13. Indeed, they are even <u>equal</u> by the corresponding Lemma for SSS Congruence V.12.1. Hence $X = C'$, and we have already shown that $\triangle ABC$ and $\triangle A'B'C'$ are similar. □

[9] The existence of the intersection point X is shown by the proposition telling equiangular triangles exist in all given sizes.

Proposition IV.7.5 (Euclid VI.6). *If two triangles have one pair of congruent angles, and the sides containing these pairs are proportional, then the triangles are similar.*

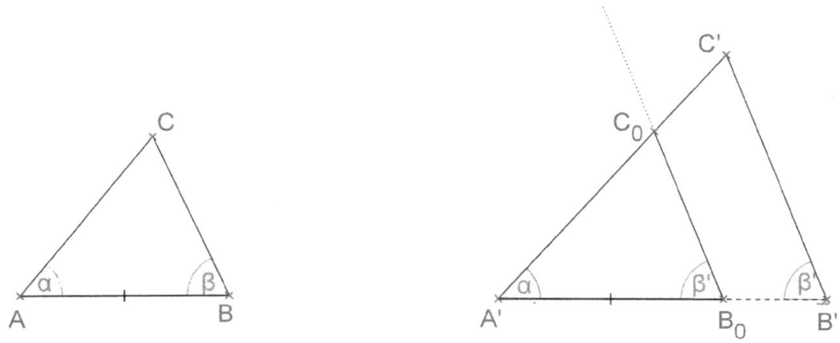

Figure IV.7.3: Addendum to the extended ASA congruence.

Proof. We use Euler's notation: angle α lies opposite to side a, angle β lies opposite to side b, and angle γ lies opposite to side c. Given are two triangles $\triangle ABC$ and $\triangle A'B'C'$ with congruent angles $\angle BAC \cong \angle B'A'C'$ and two pairs of adjacent sides b, c and b', c' which have proportional lengths:

$$\alpha = \alpha' \text{ and } \frac{b}{c} = \frac{b'}{c'}$$

Since segment length are comparable, we may assume $AB \le A'B'$, without loss of generality. There exists a point B_0 in the segment $A'B'$ such that

$$AB \cong A'B_0 \tag{IV.7.2}$$

There exists a ray $\overrightarrow{B_0C_0}$ pointing into the half plane of C' and making congruent angles $\beta' = \angle A'B'C' \cong \angle A'B_0C_0$. By congruent z-angles, the lines $B_0C_0 \parallel B'C'$ are parallel. Hence by Pasch's axiom exists the intersection point C_0 in the segment $A'C'$. For a Pythagorean plane, it is assumed the angle sum of any triangle is $180°$. Hence the triangles $\triangle A'B_0C_0$ and $\triangle A'B'C'$ are equiangular. By Euclid VI.4 (stated in proposition IV.7.2), the equiangular triangles $\triangle A'B_0C_0$ and $\triangle A'B'C'$ are similar. We obtain the proportion

$$\frac{|A'C_0|}{|A'B_0|} = \frac{|A'C'|}{|A'B'|} = \frac{b'}{c'}$$

Hence the assumption

$$\frac{b'}{c'} = \frac{b}{c} = \frac{|AC|}{|AB|}$$

and $A'B_0 \cong AB$ imply $|A'C_0| = |AC|$ and finally

$$AC \cong A'C_0 \tag{IV.7.3}$$

By equations (IV.7.2) and (IV.7.3) and $\angle CAB \cong \angle C_0AB_0$, the SAS congruence implies that the triangles $\triangle ABC$ and $\triangle AB_0C_0$ are congruent.

Since the triangles $\triangle A'B_0C_0$ and $\triangle A'B'C'$ are similar, we finally conclude the triangles $\triangle ABC$ and $\triangle A'B'C'$ are similar, as required. □

IV.7.2 Some Exercises

Problem IV.7.2. *Given are two triangles with one pair of congruent angles, and the sides across these angles are proportional to a pair of sides adjacent to the angles. Are the two triangles always similar. Give examples, counterexamples, a general reason.*

Answer. The two triangles need not always to be similar. Take two triangles for which the assumption holds in the form

$$\gamma = \gamma' \ , \quad \frac{c}{a} = \frac{c'}{a'}$$

It can happen they are not similar. Actually, this is only possible if the angle $\gamma = \gamma' < 90°$ is acute, and $\frac{c}{a} < 1$. Take the example $\gamma = \gamma' = 30°$ and $\frac{c}{a} = \frac{c'}{a'} = \frac{3}{5}$. There exist two triangles with these pieces which are not congruent, and similar neither. See the figure on page 147, it is given in the present volume, too. I call this figure a "dangling derrick." The important point is that angle γ and side c are *across* to each other.

Problem IV.7.3. *For the two triangles it is assumed*

$$\alpha = \alpha' \ , \quad \frac{b}{c} = \frac{b'}{c'}$$

Are the two triangles always similar. Give examples, counterexamples, a general reason.

Answer. The similarity has just been stated in Proposition IV.7.5. The sides b and c the ratio of which is given are adjacent to the given angle α. Hence Euclid VI.6 tells that the two triangles are similar.

Problem IV.7.4. *Draw two circles of equal radii, the center of one lying on the circumference of the other one. Draw the line through the two centers. Use some*

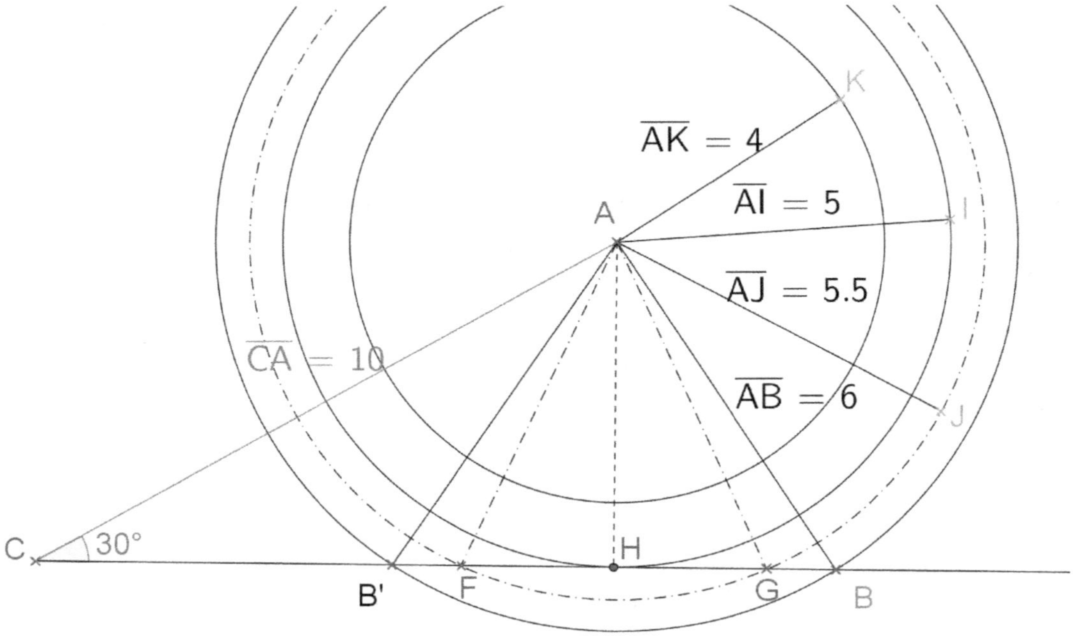

Figure IV.7.4: A Euclidean example for SSA triangle construction

of the intersection points you have just obtained, and draw two triangles with angles $30°, 30°, 120°$ *of different sizes, in two different colors.*

Let the lengths of sides of the larger triangle be a *and* c*, and the lengths of the sides of the smaller one be* a' *and* c'*. Determine the ratio* $\frac{a}{a'} = \frac{c}{c'}$*.*

Answer. The line through the two centers A and B intersects the two circles in points D and E, too. In the figure on page 148, these four points have the order $A*D*E*B$. Let C be an intersection point of the two circles. For example, we get two isosceles similar triangles $\triangle ABC \sim \triangle ACD$. Both have the base angles 30° and top angle 120°.

We determine the ratio $\frac{a}{a'} = \frac{c}{c'}$ of the sizes of the two triangles. Since $c' = a$ and $c = 3a'$, we get

$$\frac{a}{a'} = \frac{c}{c'} = \frac{3a'}{a}$$

$$\frac{a^2}{a'^2} = 3 \quad \text{and} \quad \frac{a}{a'} = \sqrt{3}$$

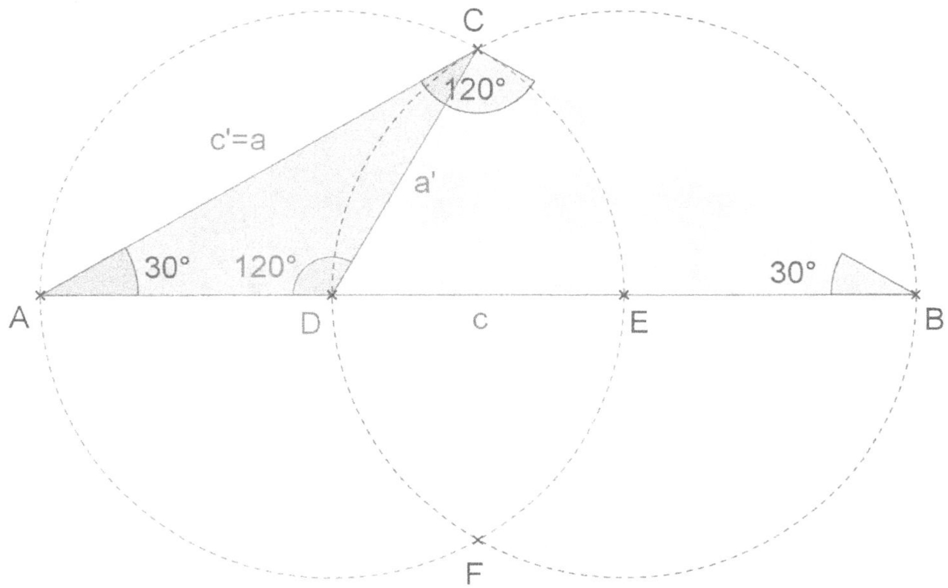

Figure IV.7.5: Two isosceles similar triangles with angles of 30° and 120°.

Problem IV.7.5. *Given is a triangle. Inscribe a square with one side part of a side of the triangle, and all vertices on the sides of the triangle.*

Answer. We erect a square onto one side a of the triangle. We connect the two new vertices of the square to the third vertex A of the triangle. The connecting lines intersect the side a in two points, which are the endpoints of one side of the square to be inscribed. Finally, we draw the inscribed square.

Problem IV.7.6. *A square is inscribed with one side lying on the side a of the triangle, and the other two vertices lying on the two remaining sides of the triangle. Determine the side x of the square in terms of the side a and the area F of the triangle.*

Answer. We drop the perpendicular from vertex A onto the opposite side. From the similar triangles $\triangle ADK \sim \triangle AFL$, we get the proportions

$$\frac{|DK|}{|AK|} = \frac{|FL|}{|AL|}$$

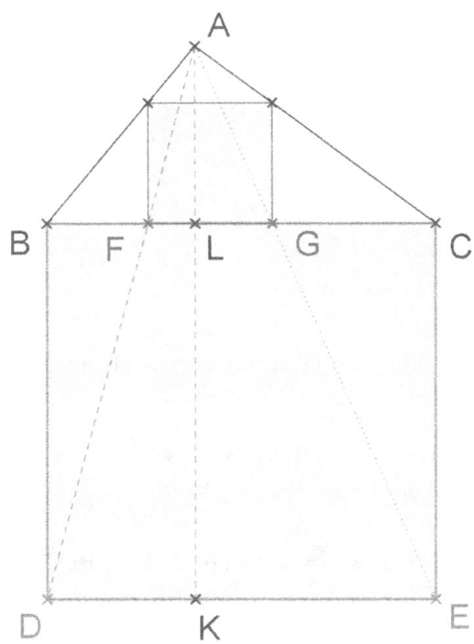

Figure IV.7.6: Inscribing a square into a triangle.

Once more one gets proportions form the similar triangles $\triangle AEK \sim \triangle AGL$, and adds

$$\frac{|EK|}{|AK|} = \frac{|GL|}{|AL|}$$
$$\frac{|DE|}{|AK|} = \frac{|FG|}{|AL|}$$
$$\frac{a}{a+h} = \frac{x}{h}$$

were h is the altitude. Hence

$$x = \frac{ah}{a+h} = \frac{2aF}{a^2+2F}$$

Problem IV.7.7. *Suppose that the endpoints of one side of the square are midpoints of two sides of the triangle. Show by dissection that the area of the square is half the area of the triangle.*

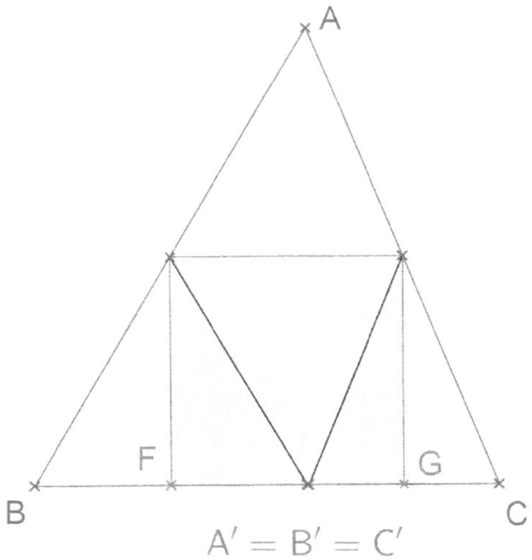

Figure IV.7.7: For $h = a$, one can fold the triangle to cover the square twice.

Problem IV.7.8. *Is the area of the square from the problem above less, more or equal to half of the area of the triangle. Give a simple folding argument to decide without any calculation.*

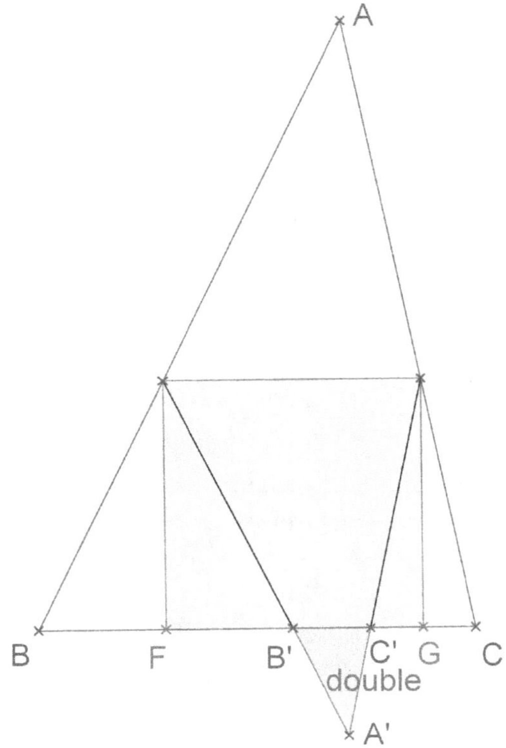

Figure IV.7.8: For $a < h$, one can fold the triangle to cover the square twice, and triangle $\triangle A'B'C'$ left over.

Answer. For the special case that the endpoints of one side of the square are midpoints of two sides of the triangle, the area of the square is half of the area of the triangle. In the figure on page 150 you see this borderline case. In all other cases, the area of the square is less than half of the area of the triangle.To see this, we fold the triangle along three sides of the square. We get three reflection images A', B' and C' of the vertices of the triangle. Because of the right angles of the square, the folded sides touch along the lines $A'B'$ and $A'C'$. In the figure on pages 151, you see the case that triangle side $a < h$ is less than the altitude h, and in the figure on page 152, the case that triangle side $a > h$ is longer than the altitude. In the first case, one covers the square and an extra triangle $\triangle A'B'C'$ is left over in the opposite half plane. In the second case, one again covers the square, and the part of the extra triangle $\triangle A'B'C'$ inside the square is covered twice.

Problem IV.7.9. *Given are two intersecting lines l and m and a point A different*

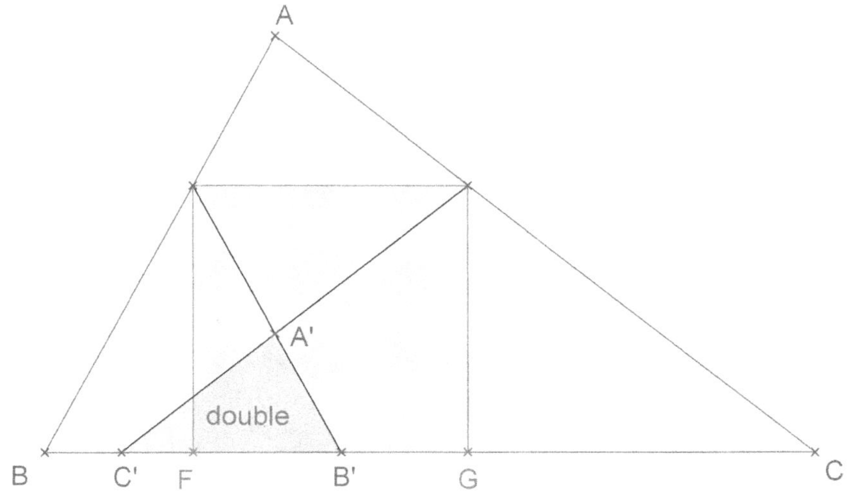

Figure IV.7.9: For $a > h$, one can fold the triangle to cover the square twice, and part of triangle $\triangle A'B'C'$ is covered twice.

from their intersection point. We have to construct a circle through the point A touching both lines. Describe the construction given in the figure on page 153.

Answer.

- We draw the angular bisector of the the two given lines, bisecting the angle α in the interior of which the given point A lies.

- We choose any point O' on the bisector and draw a circle around it touching both line l and line m.

- We draw the ray emanating from the intersection point Z of the two given lines and pointing into the interior of the angle α. This ray intersects the circle in two points C and D.

- The parallel to line DO' through point A intersects the bisector in point O. The circle around O through point A does touch the two given lines and solves the problem.

- A second solution is obtained by drawing the parallel to line CO' through point A. This line intersects the bisector in the center O_2. The circle around O_2 through point A does touch the two given lines and yields the second solution of the problem, which has not been drawn.

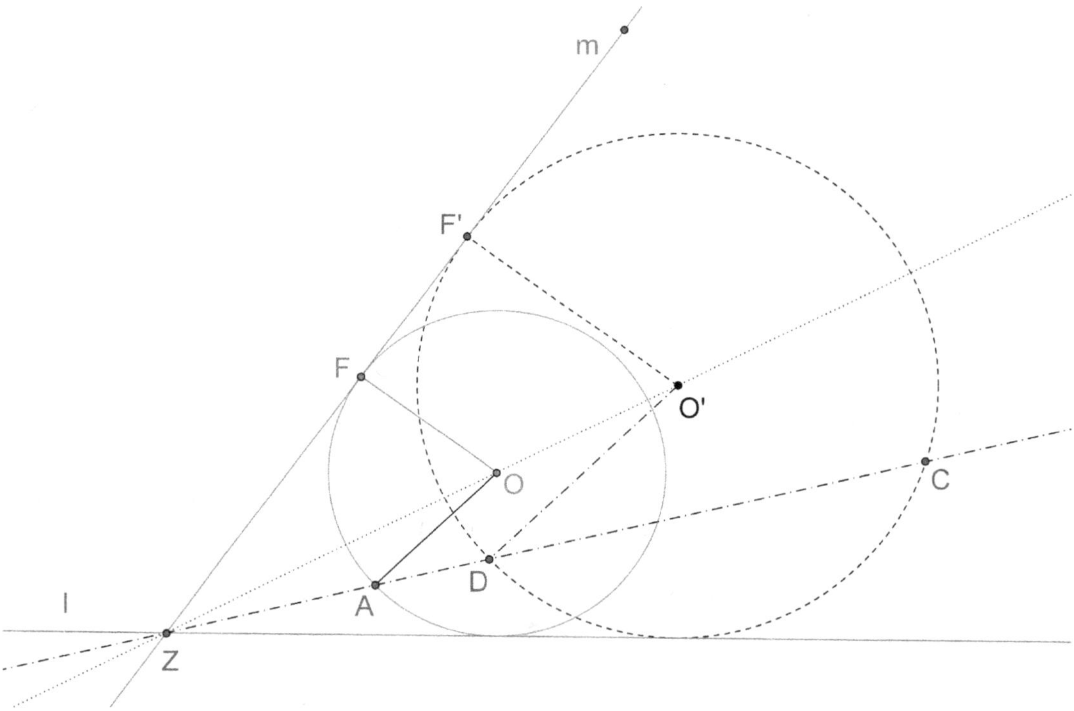

Figure IV.7.10: Constructing a circle through a point and touching two lines.

Reason for the construction. The two triangles $\triangle ZAO \sim \triangle ZDO'$ are equiangular because the lines AO and DO' are parallel by construction. Because the lines FO and $F'O'$ are both perpendicular to the same line m and hence parallel, the triangles $\triangle ZFO \sim \triangle ZF'O'$ are equiangular . Since equiangular triangles are similar, we get the proportions

$$\frac{ZO}{AO} = \frac{ZO'}{DO'} \quad \text{and} \quad \frac{ZO}{FO} = \frac{ZO'}{F'O'}$$

Division yields

$$\frac{AO}{FO} = \frac{DO'}{F'O'} = 1$$

Indeed, the latter quotient equals 1 since both points D and F' lie on a circle around O' . Hence the former quotient equals one, too. Hence both points A and F lie on a circle $\mathcal{C}$ around O. This means that the constructed circle $\mathcal{C}$ touches the line m. Since the centers O' and O lie on the angle bisector of the two given lines l and m, the circle $\mathcal{O}$ touches the line l, too. □

Problem IV.7.10. *Given are two points A and B and a line l. Construct a circle through the two points touching the line.*

First ideas. It is useful to consider these cases:

(a) One of the points lies on the line.

(b) The lines AB and l are different and parallel.

(c) The lines AB and l intersect perpendicularly.

(d) The lines AB and l intersect, at any other angle.

We have solved case (a) earlier in Problem I.2.7 in the section "A simplified axiomatic system of geometry". I leave cases (a) and (b) to the reader.

Question. What assumption about the two points and the line is needed that the problem can be solvable is cases (c) and (d).

Answer. The two points A and B need to lie on the same side of line l, or one of them on the line. The most interesting case (d) is explained below.

□

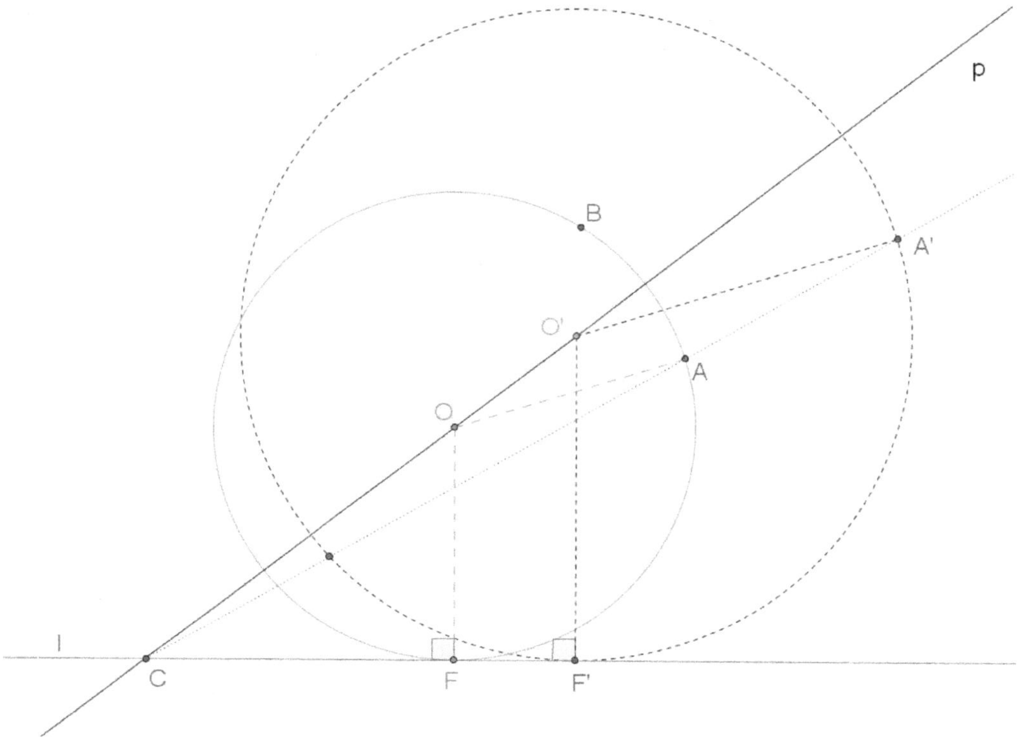

Figure IV.7.11: Constructing a circle through two points touching a line.

Problem IV.7.11. *Given is a line l, and two points A and B on the same side of this line. It is assumed that the lines AB and l intersect, but not perpendicularly. Construct a circle through the two points touching the line.*

Construction for Problem IV.7.11. Let p the perpendicular bisector of segment AB. The lines p and l intersect, say at point C. We use point C as a center for similarities.

We choose any point $O' \neq C$ on the line p which lies on the same side as points A and B. We draw the circle $\mathcal{D}'$ around O' touching the line l, say in point F'. This circle can be mapped by a similarity with center C to the circle to be constructed.

We choose an intersection point A' of $\overrightarrow{CA} \cap \mathcal{D}'$. The parallel of line $A'O'$ intersects the perpendicular bisector p in point O. We construct the circle $\mathcal{D}$ around O touching the line l in point F. □

Complete justification for Problem IV.7.11.

Question. Why do the two rays $\overrightarrow{CA}$ and $\overrightarrow{CB}$ both intersect the circle $\mathcal{D}'$.

Answer. One of the points A and B lies in the interior of the angle $\angle O'CF'$, say point A. By the Crossbar Theorem [10] the segment $O'F'$ and the ray $\overrightarrow{CA}$ intersect, say in point Q.

The point Q is in the interior of circle $\mathcal{D}'$, hence the line CA intersects the circle $\mathcal{D}'$. Since all points of a circle lie on the same side of the tangent as the center, even the ray $\overrightarrow{CA}$ intersects the circle $\mathcal{D}'$, indeed in two points. These are rather obvious consequences of the circle-line intersection property, see Proposition V.12.21 and Proposition V.12.18 from the section of neutral geometry of circles and continuity.

The two rays $\overrightarrow{CA}$ and $\overrightarrow{CB}$ are images obtained from each other by reflection across the line p. The circle $\mathcal{D}'$. is symmetric to this line. Hence the ray $\overrightarrow{CB}$ intersects the circle $\mathcal{D}'$, indeed in two further points.

Question. By means of similar triangles, check that circle $\mathcal{D}$ goes through point A.

Answer. The triangles $\triangle COA$ and $\triangle CO'A'$ are equiangular by construction, and hence similar:

$$\frac{|CO|}{|OA|} = \frac{|C'O'|}{|O'A'|}$$

The right triangles $\triangle COF$ and $\triangle CO'F'$ are equiangular by construction, and hence similar:

$$\frac{|CO|}{|OF|} = \frac{|C'O'|}{|O'F'|}$$

Hence

$$\frac{|OA|}{|OF|} = \frac{|O'A'|}{|O'F'|} = 1$$

by calculation, and since A' and F' lie on a circle around O'. Hence A and F lie on a circle around O, as to be shown.

□

IV.7.3 Secants in a Circle

Theorem IV.7.3 (Theorem of chords (Euclid III.35)). *If two chords cut each other inside a circle, the product of the segments on one chord equals the product of the segments on the other chord.*

Assume two chords intersect each other outside a circle. The product of the segments, measured from the point of intersection to the two intersection points with the circle are equal for both chords.

[10] A segment with endpoints on the two sides of an angle and a ray emanating from its vertex into the interior of the angle intersect. See the section about the axioms of order, Proposition V.12.1

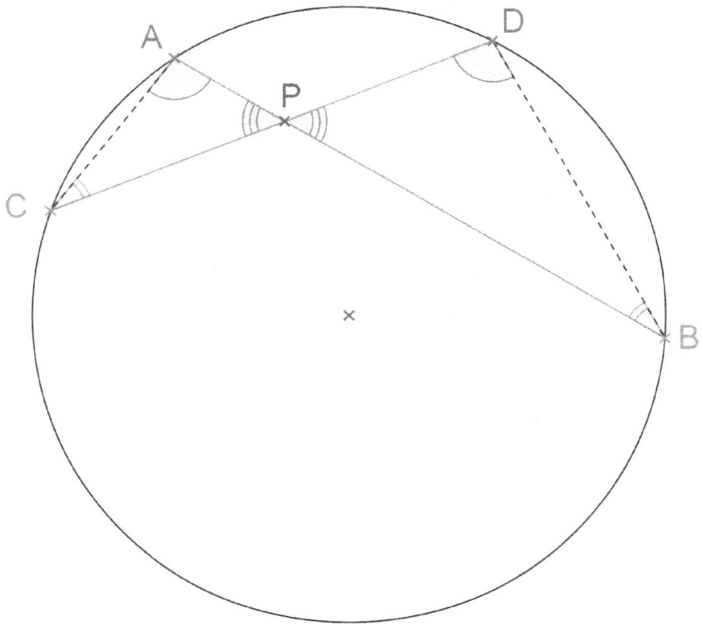

Figure IV.7.12: Theorem of chords, case of chords intersecting inside the circle

Problem IV.7.12. *Provide a drawing, with appropriate notation.*

Proof using proportions. Let AB and CD be two chords of circle γ intersecting at point P inside the circle. We have to check whether $|PA| \cdot |PB| = |PC| \cdot |PD|$. Following Legendre, I am using proportions. Equiangular triangles are, by definition, triangles with pairwise congruent angles at their corresponding vertices. Congruent circumference angles, as shown in Euclid III.21, occur at vertices A and D, as well as at vertices C and D. Furthermore, we get congruent vertical angles at vertex P. The vertices of the equiangular triangles have to be listed in such an order that these congruent angles occur at corresponding vertices. Hence we get the equiangular triangles

$$\triangle PAC \sim \triangle PDB$$

By Euclid VI.4

```
The sides of equiangular triangles are proportional.
```

Hence the ratios of corresponding sides are the same for two equiangular triangles.

$$\frac{|PA|}{|PC|} = \frac{|PD|}{|PB|} \tag{IV.7.4}$$

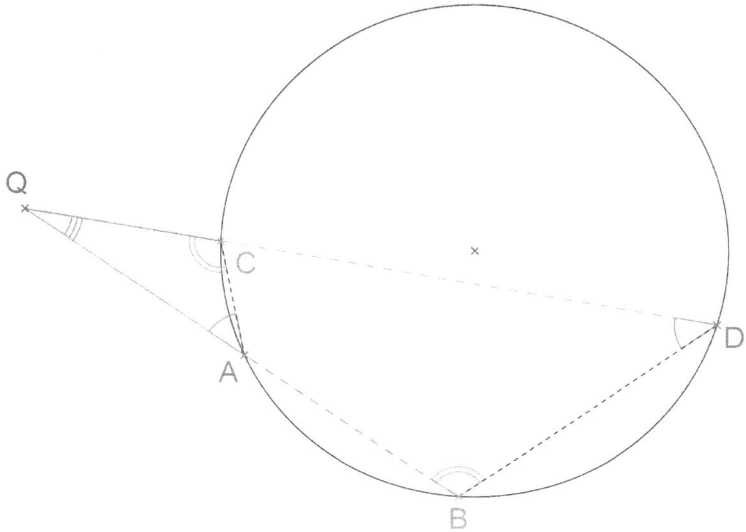

Figure IV.7.13: Theorem of chords, case of chords intersecting outside the circle

Now multiplying with the denominators yields

$$|PA| \cdot |PB| = |PC| \cdot |PD|$$

which is just the claim of Euclid III.35.

The proof for the case of the segments intersecting outside the circle is almost identical. But note that the two segments QA and QB now overlap. □

For the case where the extension of the chords intersect outside the circle, one obtains a further result by considering the tangent as a limit of a small secant. In the figures on page 159, we have illustrated the limit $C \to T, D \to T$, where T is the touching point of a tangent drawn from point Q to the circle. Thus one is lead to claim Euclid's next result.

Theorem IV.7.4 (Theorem of chord and tangent (Euclid III.36)). *From a point outside a circle, a tangent and a second are drawn. The square of the tangent segment equals the product of the segments on the chord, measured from the point outside to the two intersection points with the circle.*

Problem IV.7.13. *Provide a drawing, with appropriate notation.*

Independent proof, again using proportions. Let Q be a point outside circle γ, let T be the touching point of the tangent to the circle through Q, and let AB be a chord

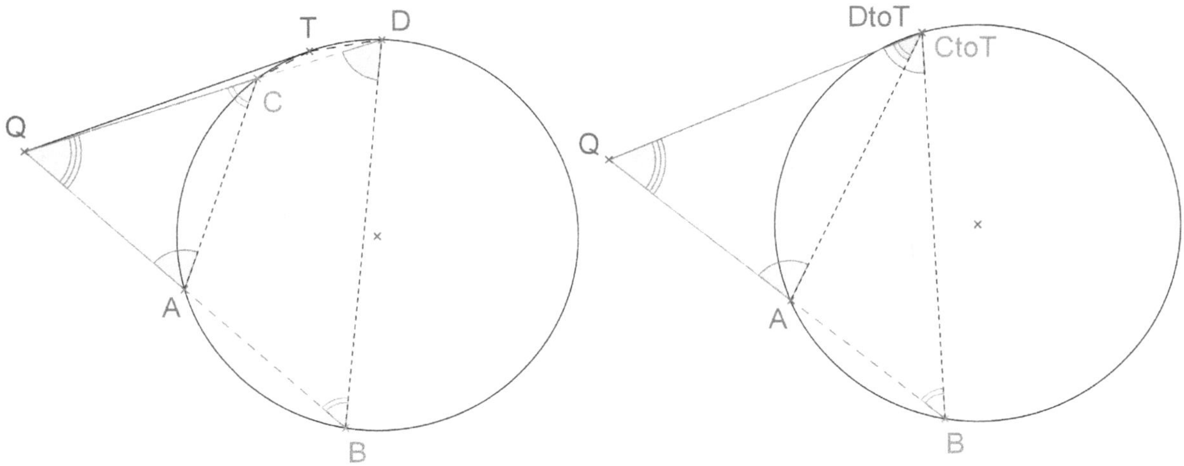

Figure IV.7.14: Getting the limit $C \to T, D \to T$

of the circle the extension of which runs through Q. We assume that B lies between Q and A.

I have to check whether $|QT|^2 = |QA| \cdot |QB|$. To this end, one compares the triangles $\triangle QAT$ and $\triangle QTB$. We use Euclid III.32:

> `The angle between a tangent line and a chord is congruent to the circumference angle of the arc corresponding to the chord.`

Hence the angles at vertices A and T, for the two triangles, respectively, are congruent. With vertices listed in an order that these congruent angles occur at corresponding vertices, we get equiangular triangles $\triangle QAT \sim \triangle QTB$. By Euclid VI.4:

> `The sides of equiangular triangles are proportional.`

Hence the ratios of corresponding sides are the same for two equiangular triangles.

$$\frac{|QA|}{|QT|} = \frac{|QT|}{|QB|} \tag{IV.7.5}$$

and multiplying with the denominators yields the result. □

IV.7.4 The Lens Equation

Proposition IV.7.6. *In a trapezoid $\square ABCD$ is drawn a third parallel to the two parallel sides through the intersection point S of the diagonals. The trapezoid sides cut congruent segments $f = FS$ and $g = SG$ out of the third parallel.*

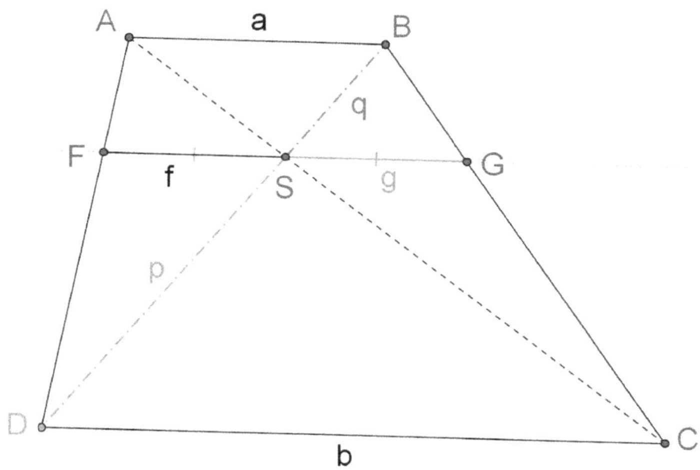

Figure IV.7.15: A trapezoid and the lens equation.

(i) *The segments $f = FS$ and $g = SG$ cut out on the third parallel are congruent.*

(ii) *The lengths of the parallel sides $a = AB$ and $b = CD$ and the segment $f = FS$ satisfy the lens equation*

$$\frac{1}{a} + \frac{1}{b} = \frac{1}{f} \tag{IV.7.6}$$

Proof. We introduce the segments $p = DS$ and $q = SB$. Now we use similar triangles to get three equations among the lengths a, b, f, g, p, q

Question. Use two similar triangles with common vertex D to find a proportion among these quantities.

Answer. The triangles $\triangle DSF$ and $\triangle DBA$ are equiangular and hence similar. One gets the proportion

$$\frac{p}{f} = \frac{p+q}{a} \tag{IV.7.7}$$

Question. Use two similar triangles with common vertex B to find a proportion among these quantities.

Answer. The triangles $\triangle BSG$ and $\triangle BDC$ are equiangular and hence similar. One gets the proportion

$$\frac{q}{g} = \frac{p+q}{b} \tag{IV.7.8}$$

Question. Use two similar triangles with common vertex S to find a proportion among these quantities.

Answer. The triangles $\triangle SAB$ and $\triangle SCD$ are equiangular and hence similar. One gets the proportion

$$\frac{a}{q} = \frac{b}{p} \tag{IV.7.9}$$

We need now to eliminate p and q from these three equation. Dividing equation (IV.7.7) by equation (IV.7.8) yields

$$\frac{p \cdot g}{f \cdot q} = \frac{b}{a}$$

Together with equation (IV.7.9) one gets

$$\frac{p \cdot g}{f \cdot q} = \frac{p}{q}$$

and hence $f = g$ as required.

Question. Use this result now to get the lens equation out of the two equations (IV.7.7) and (IV.7.8).

Answer. We add these two equations and divides by $p + q$ to obtain the lens equation (IV.7.6).

□

The trapezoid from the figure on page 160 appears in the context of geometric optics. Recall that the perpendicular to the plane of the lens through its center is called the *optical axis*. For the light rays through a thin lens and close to the optical axis, one can empirically gets to the following three observations:

(a) Light rays parallel to the optical axis coming in,—say from the left—are bundled at the focus G behind the lens.

(b) There is a second focus F on the other side of the lens at the same distance. Light rays starting at focus F are bend on the other side of the lens into rays parallel to optical axis.

(c) Light rays,, even those coming in from different directions, go straight through the center of the lens.

These assertions are in reality only an approximation, but they are exact enough for a thin lens to be useful. These assumption lead to a trapezoid, which has the unusual form shown in the figure on page 162. The third parallel FG from above turns out to

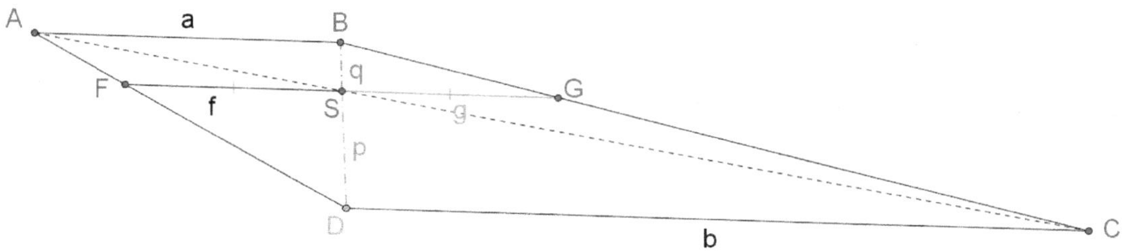

Figure IV.7.16: The trapezoid appearing for an optical lens.

be the optical axis. The light source is located at vertex A, the lens lies in the plane BD. The light rays emanated point A are focused at vertex C. We can check this to true at least for three rays:

- the ray AD through the left focus F, which is bend to the ray CD parallel to the optical axis;
- the ray AB parallel to the optical axis, which is bend to the ray BC through the right focus G;
- the ray ASC which goes straight through the center S of the lens.

Too, Proposition IV.7.6 now confirms the well-known lens equation (IV.7.6).

Problem IV.7.14. *A candle is standing at distance* $a = 3$ *from a lens with focus length* $f = 2$. *At which distance* b *behind the lens has one to put a screen to get a sharp image of the candle flame. Give a construction and calculate the distance* b.

IV.8 The Pythagorean and Related Theorems

IV.8.1 The Leg and the Altitude Theorems

Problem IV.8.1. *The figure on page 163 explains the common definitions for the lengths* a, b, c *and* p, q, h *from a right triangle. Find an example where all six segments have lengths with of integer values.*

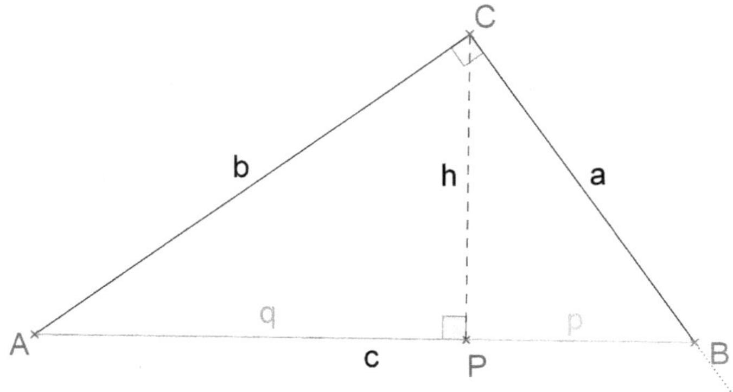

Figure IV.8.1: The common notation for a right triangle

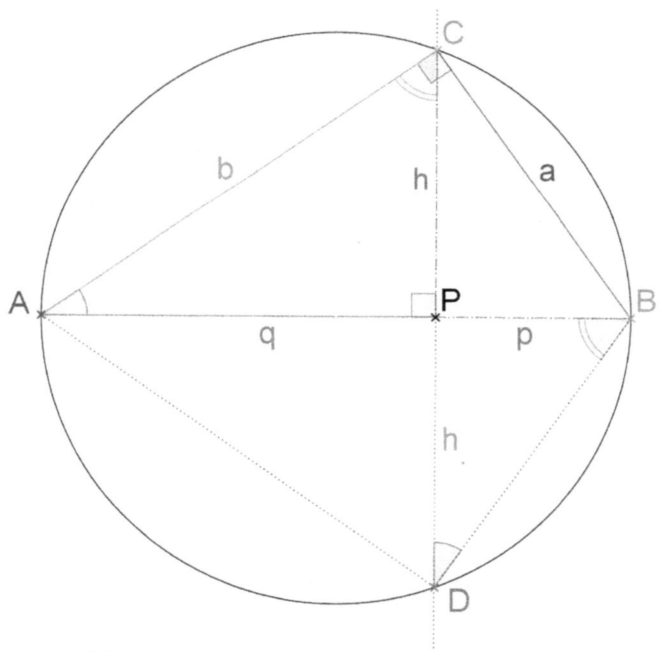

Figure IV.8.2: The altitude theorem for a right triangle

Answer. The smallest example, I can find is

$$a = 15\,,\ b = 20\,,\ c = 25 \quad \text{and} \quad p = 9\,,\ q = 16\,,\ h = 12$$

Problem IV.8.2. *Explain for which special situation Euclid III.35 implies the theorem about the altitude of a right triangle, usually stated as* $h^2 = pq$.

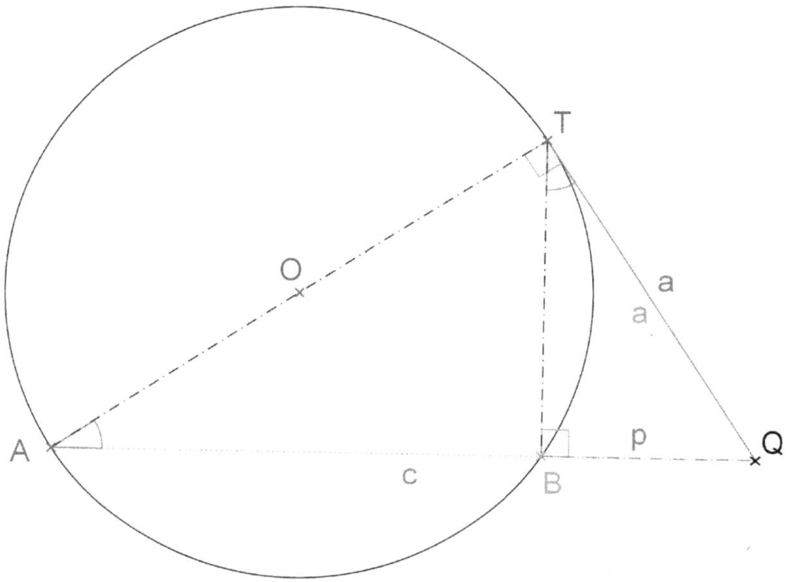

Figure IV.8.3: The leg theorem

Answer. Take for segment AB a diameter of circle γ and chose CD to be any segment perpendicular to that diameter. One gets a right $\triangle ABC$, with P as foot point of its altitude. Because of $PC \cong PD$ and $|PC| = |PD|$, Euclid III.35 implies $pq = |PA| \cdot |PB| = |PC| \cdot |PD| = h^2$. □

Problem IV.8.3. *Explain how Euclid III.36 implies the leg theorem $a^2 = pc$.*

Answer. In Euclid III.36, as given in theorem IV.7.4, we choose points A and T on a diameter of circle γ, and point Q on the tangent to the circle at point T. We see that $\triangle ATQ$ is a right triangle, because tangent and radius are perpendicular to each other. The segment AQ intersects the circle in a second point B. The triangle $\triangle ATQ$ has altitude TB, because Thales' theorem shows that the angle $\angle ABT$ is right. Now Euclid III.36 tells that the square of one leg QT equals the product of the hypothenuse time the projection of that leg onto the hypothenuse. After renaming $T \to C, B \to P, Q \to B, A \to A$, one gets the statement in its usual form $a^2 = pc$. □

IV.8.2 The Pythagorean Theorem

Problem IV.8.4. *Explain how Euclid III.36 implies the Pythagorean Theorem $a^2 + b^2 = c^2$.*

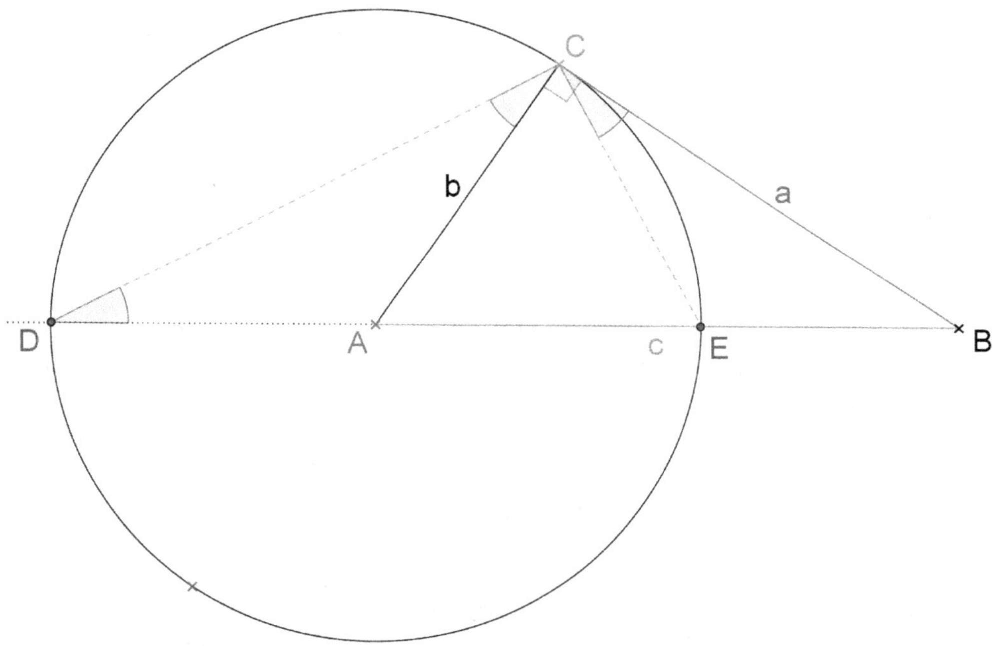

Figure IV.8.4: The Pythagorean Theorem

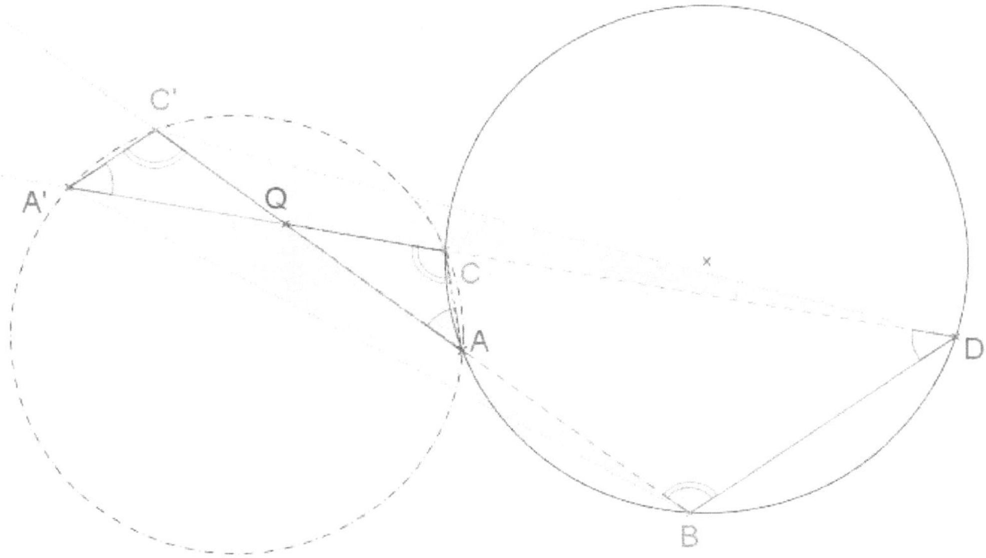

Figure IV.8.5: A pair of triangles $\triangle A'QB$ and $\triangle C'QD$ of equal area directly confirms the theorem of chords

Answer. For the given right triangle $\triangle ABC$, we draw circle with center A through point C. We draw the tangent at point C and see that point B lies on the tangent since tangent and radius are perpendicular to each other.

We draw the ray $\overrightarrow{BA}$. Let points E and D be the endpoints of the diameter on this ray. Now Euclid III.36, as given in theorem IV.7.4, tells that

$$|BC|^2 = |BD| \cdot |BE|$$

In terms of the sides of the triangle $\triangle ABC$, this shows that

$$a^2 = (c+b)(c-b) = c^2 - b^2 \quad \text{and } a^2 + b^2 = c^2$$

We got the Pythagorean Theorem. □

Remark. In the spirit of Euclid, one has to understand the product of segments occurring in the theorems above as *areas of rectangles or squares.* The equality of their areas can really be shown by obtaining one from the other in a finite sequence of cuts and pastes. In the figure below, I want to visualize this interpretation. Take the case of chords intersecting at point Q outside the circle. We extend the two chords to the other side of their intersection point and transfer the segments QA and

QC to these opposite rays to obtain a quadrilateral $\square A'ACC'$ with axial symmetry. It is not hard to check that this quadrilateral has a circum circle and the triangles $\triangle QAC \cong \triangle QA'C'$ are congruent.

The two circle lemma implies that the chords $A'C' \parallel BD$ are parallel. Hence we have obtained a new pair of similar triangles

$$\triangle QA'C' \sim \triangle PDB$$

which have the center of similarity Q.

The figure also contains two triangles of equal area, which confirm Euclid's theorem III.35 directly as an *equality among areas*. Since $A'C' \parallel BD$, the triangles

$$|\triangle A'C'B| \doteq |\triangle A'C'D|$$

have equal base $A'C'$ and equal height, and hence equal areas. Their intersection is the triangle $\triangle QA'C'$. We subtract this triangle from both sides, and obtain the pair of triangles

$$|\triangle A'QB| \doteq |\triangle C'QD|$$

again of equal areas. Now the claim of Euclid III.35—

$$|QA'| \cdot |QB| = |QC'| \cdot |QD|$$

can easily be obtained as an equality among areas.

IV.8.3 Some Number Theory Inspired by Pythagoras' Theorem

Problem IV.8.5. *In the figure on page 163 is shown the common notation for the six segments a, b, c and p, q, h which occur in a right triangle. Any two among the six segments a, b, c and p, q, h determine the other four ones. We assume that the hypothenuse c and the altitude h are given. Find explicit formulas for the legs a and b and the projections p and q in terms of the hypothenuse c and the altitude h. To determine the sign of the roots occurring in the formulas, we assume that $a \geq b$.*

Answer. From the Pythagorean theorem $a^2 + b^2 = c^2$ and the formula $ab = ch$, one obtains $a + b$.

$$(a+b)^2 = a^2 + b^2 + 2ab = c^2 + 2ch$$
$$a + b = \sqrt{c^2 + 2ch}$$

From known product ab and sum $a+b$, one gets a quadratic equation for the unknown a. Thus one determines a and finally b. The sign of the roots is determined by the assumption that $a \geq b$.

$$a^2 - a\sqrt{c^2+2ch} + ch = 0$$

$$a = \frac{\sqrt{c^2+2ch} + \sqrt{c^2+2ch-4ch}}{2} = \frac{\sqrt{c^2+2ch} + \sqrt{c^2-2ch}}{2}$$

$$b = \frac{\sqrt{c^2+2ch} - \sqrt{c^2-2ch}}{2}$$

Now p and q can be calculated by the leg theorem.

$$p = \frac{a^2}{c} = \frac{\left(\sqrt{c^2+2ch} + \sqrt{c^2-2ch}\right)^2}{4c} = \frac{2c^2 + 2\sqrt{(c^2+2ch)(c^2-2ch)}}{4c}$$

$$= \frac{2c^2 + 2\sqrt{c^4-4c^2h^2}}{4c} = \frac{c + \sqrt{c^2-4h^2}}{2} \quad \text{and}$$

$$q = \frac{c - \sqrt{c^2-4h^2}}{2}$$

Lemma IV.8.1. *For any solution of the equation $a^2+b^2=c^2$ with natural numbers a,b,c it is impossible than a and b are both odd.*

Proof. Assume towards a contradiction that a and b are both odd and $a^2+b^2=c^2$. There exists a natural number $n \geq 0$ such that $a = 2n+1$. One checks that $a^2 = (2n+1)^2 = 4(n^2+n)+1 \equiv 1 \mod 4$. Similarly one obtains $b^2 \equiv 1 \mod 4$. Hence $c^2 \equiv 2 \mod 4$ and c is even, but c^2 is not divisible by 4, which is impossible. □

Proposition IV.8.1. *All solutions of the equation $a^2+b^2=c^2$ with natural numbers a,b,c, for which b divisible by a* higher *prime power of 2 than a, are given by the formulas*

$$a = g(u^2-v^2)\,, \quad b = 2guv\,, \quad c = g(u^2+v^2) \tag{IV.8.1}$$

with integer $g \geq 1$ and $u > v \geq 1$. Moreover, if u and v are relatively prime and $u+v$ is odd, then $g = \gcd(a,b,c) = \gcd(a,b) = \gcd(a,c) = \gcd(b,c)$.

Proof. It is enough to deal with the case that $\gcd(a,b,c) = 1$. Hence $\gcd(a,b) = \gcd(a,c) = \gcd(b,c) = 1$ holds, too. By lemma IV.8.1 one concludes that b is even, and a and c are odd. Hence $b+c$ and $c-b$ are both odd and $\gcd(c+b,c-b) = 1$. Because of

$$(c+b)(c-b) = c^2 - b^2 = a^2$$

and the uniqueness of the prime decomposition, we conclude that the numbers $c+b$ and $c-b$ are both perfect squares. We conclude

$$c+b=m^2\ ,\quad c-b=n^2 \quad \text{and} \quad a=m\cdot n$$

where $m>n$ are both odd natural numbers. Define the natural numbers

$$u:=\frac{m+n}{2} \quad \text{and} \quad v:=\frac{m-n}{2}$$

Clearly $u>v$ are both natural numbers, they are relatively prime and $u+v=m$ is odd. We easily check that

$$\begin{aligned} b &= \frac{m^2-n^2}{2} = 2uv \\ c &= \frac{m^2+n^2}{2} = u^2+v^2\ , \quad \text{and finally} \\ a &= m\cdot n = u^2-v^2 \end{aligned}$$

to confirm that equations (IV.8.1) hold with $g=1$, as to be shown. □

Corollary 15. *The unit circle* $\{(x,y) : x^2=y^2=1\}$ *contains infinitely many points for which both coordinates are rational.*

Reason. One puts

$$x=\frac{u^2-v^2}{u^2+v^2} \quad \text{and} \quad y=\frac{2uv}{u^2+v^2}$$

with any integers $u,v\in\mathbf{Z}$. □

Remark. Using complex numbers, the equations (IV.8.1) can be put together as

$$a+ib=g(u+iv)^2$$

Hence one gets once more

$$c^2=(a+ib)(a-ib)=g^2(u+iv)^2(u-iv)^2=g^2(u^2+v^2)^2$$
$$\text{and} \quad c=g(u^2+v^2)$$

Corollary 16. *In the ring* $\mathbf{Z}+i\mathbf{Z}$ *of Gaussian integers, the number* $a+ib$ *is a perfect square if and only if* a^2+b^2 *is a perfect square.*

Problem IV.8.6. *Find all cases for which all six segments* a,b,c *and* p,q,h *have integer values, and* $\gcd(a,b,c,p,q,h)=1$. *Give formulas for these six segments in terms of integer solutions of* $a'^2+b'^2=c'^2$ *with* $\gcd(a',b')=1$.

Proof. These integer solutions are

$$\text{(IV.8.2)}\qquad \begin{aligned} a = a' \cdot c' \ , \quad b = b' \cdot c' \ , \quad c = c'^2 \ , \\ p = a'^2 \ , \quad q = b'^2 \ , \quad h = a' \cdot b' \end{aligned}$$

Here is a simple reason. From proposition IV.8.1 we know that $a = a' \cdot g \ , \quad b = b' \cdot g \ , \quad c = c' \cdot g$ with $g = \gcd(a,c) = \gcd(b,c)$. From $ab = ch$ we get

$$h = \frac{ab}{c} = \frac{(a'b')g}{c'} = a'b' \cdot \frac{g}{c'}$$

which is assumed to be an integer. Now $\gcd(a',c') = 1$ and $\gcd(b',c') = 1$ imply $\gcd(a'b',c') = 1$. But c' divides the product $(a'b') \cdot g$. Hence Euclid's lemma implies than c' divides g. In other words, $\frac{g}{c'}$ is an integer. The remaining lengths are

$$p = \frac{a^2}{c} = a'^2 \cdot \frac{g}{c'} \quad \text{and} \quad q = b'^2 \cdot \frac{g}{c'}$$

We see that both are divisible by the integer $\frac{g}{c'}$, as are the lengths h as well as a and b. Hence the assumption $\gcd(a,b,c,p,q,h) = 1$ implies that $g = c'$. Finally, one obtains the formulas (IV.8.2). □

Corollary 17. *In the case that all six segments a, b, c and p, q, h have integer values, and $\gcd(a,b,c,p,q,h) = 1$, all segments c, p, q and $c + 2h$ and $c - 2h$ are perfect squares.*

Reason. One see directly from the equations (IV.8.2) that c, p, q are perfect squares. Moreover

$$c + 2h = c'^2 + 2a' \cdot b' = a'^2 + b'^2 + 2a' \cdot b' = (a' + b')^2$$
$$\text{and} \quad c - 2h = (a' - b')^2$$

□

IV.8.4 Some (Sandbox) Applications

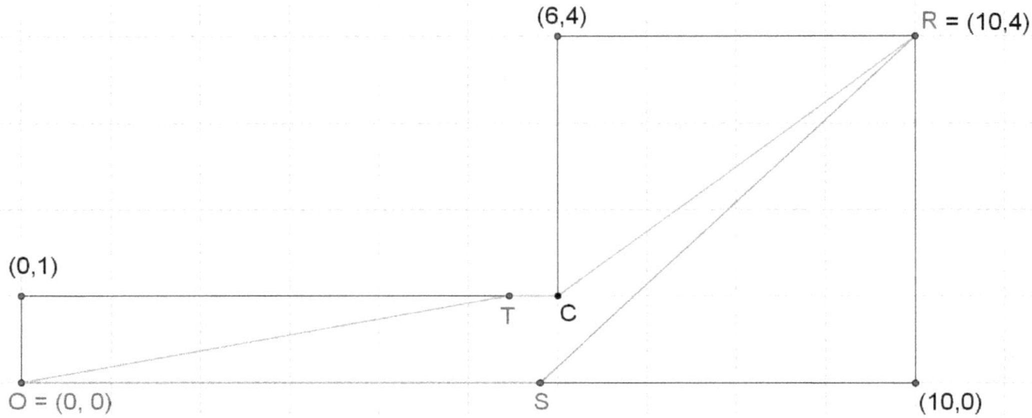

Figure IV.8.6: Two equally good runners start at O and R. Do they better meet at point S or point T.

Problem IV.8.7. *Two equally fast runners start at the opposite points O and R of the place shown in the figure on page 171. They are allowed to run across the place, but cannot enter any space outside the place. They want to meet on the boundary. Can they meet quicker at point S or at point T.*

(i) *Calculate the distance $|OS|$ for which $|OS| = |SR|$.*

(ii) *Calculate the point T such that $|OT| = |TC| + |CR|$.*

(iii) *Calculate the distance $|OT|$ and decide whether which distance is shorter $|OT|$ or $|OS|$.*

Answer. **(i)** Let $S = (x, 0)$ be the coordinates of the meeting point S. Since the distances $|OS| = |SR|$ are equal, one calculates

$$\begin{aligned} x &= \sqrt{(10-x)^2 + 4^2} \\ x^2 &= 100 - 20x + x^2 + 16 \\ x &= 5.8 \end{aligned}$$

(ii) Let the distance $|OT| = x$. Hence the coordinates of the meeting point are $T = (\sqrt{x^2-1}, 1)$. Since the distances $|OT| = |TC| + |CR|$ are equal, $|TC| =$

$6 - \sqrt{x^2 - 1}$ and $|CR| = \sqrt{3^2 + 4^2} = 5$ gives

$$x = 11 - \sqrt{x^2 - 1}$$
$$x^2 - 1 = (11 - x)^2 = 121 - 22x + x^2$$
$$x = \frac{122}{22} = 5 + \frac{6}{11}$$

(iii) We see that point T is where to meet quicker.

Remark. With $T = (t, 1)$

$$\sqrt{t^2 + 1} = 6 - t + \sqrt{3^2 + 4^2} = 11 - t$$
$$t^2 + 1 = 121 - 22t + t^2$$
$$t = 5 + \frac{5}{11}$$

We get the same distance

$$|OT| = |TC| + |CR| = (6 - t) + 5 = 5 + \frac{6}{11}$$

IV.8.5 The Parallelogram Equation

Proposition IV.8.2. *The sum of the squares of the four sides of a parallelogram equals the sum of the squares of the diagonals.*

Remark. Any real numbers satisfy

$$(a + b)^2 + (b - a)^2 = 2a^2 + 2b^2$$

The corresponding equation holds for vectors in a space with an inner product. For a parallelogram, the vectors along the sides and diagonals are

$$a := \overrightarrow{AB}\,, \quad -a := \overrightarrow{CD}\,,$$
$$b := \overrightarrow{BC} = \overrightarrow{AD}$$
$$a + b = \overrightarrow{AC}\,, \quad b - a = \overrightarrow{BD}$$

In a vector space with an inner product, the corresponding identity

$$(a + b)^2 + (b - a)^2 = 2a^2 + 2b^2$$

implies that the sum of the squares of the diagonals of a parallelogram equals twice the sum of the squares of two adjacent sides. In this way, we have an analytic proof of the parallelogram equation.

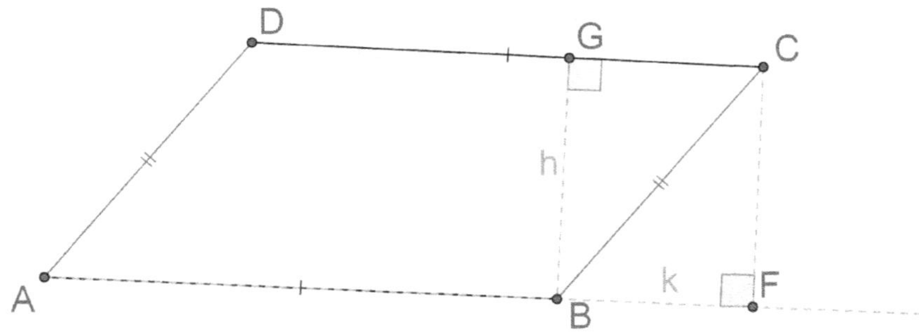

Figure IV.8.7: Proof of the parallelogram equation.

Problem IV.8.8 (Parallelogram equation). *The sum of the squares of the diagonals of a parallelogram equals the sum of the squares of its four sides. Prove the parallelogram equation*

$$|AC|^2 + |BD|^2 = 2|AB|^2 + 2|BC|^2$$

using the case and the notation of the figure on page 173.

Proof. For the parallelogram $\square ABCD$, we may assume $a = |AB| > |BC| = b$, and the angle at B to be obtuse. We drop the perpendiculars from vertex C onto the extension of side AB, and from vertex B onto side CD. The foot points are F and G, respectively. Let the rectangle $\square BFCG$ have side lengths $|BG| = |CF| = h$ and $|BF| = |CG| = k$. Apply Pythagoras' Theorem three times:

$$\begin{aligned} |AC|^2 &= |AF|^2 + |FC|^2 = (a+k)^2 + h^2 && \text{for triangle } \triangle AFC; \\ |BD|^2 &= |BG|^2 + |GD|^2 = h^2 + (a-k)^2 && \text{for triangle } \triangle BGD; \\ h^2 + k^2 &= |BF|^2 + |FC|^2 = |BC|^2 = b^2 && \text{for triangle } \triangle BFC. \end{aligned}$$

Adding the first two equations yields

$$|AC|^2 + |BD|^2 = (a+k)^2 + (a-k)^2 + 2h^2 = 2a^2 + 2k^2 + 2h^2 = 2a^2 + 2b^2$$

We see that the sum of the squares of the diagonals of parallelogram $\square ABCD$ equals the sum of the squares of its four sides. $\square$

IV.9 Trigonometry

IV.9.1 The Law of Sines

Definition IV.9.1 (sin function for acute and obtuse angles). The ratio $\frac{a}{c}$ of the leg a across the acute angle α to the hypothenuse of a right triangle is the *sin function* of this angle:

$$\sin \alpha = \frac{a}{c}$$

For right and acute angles, one defines

$$\sin(180^\circ - \alpha) = \sin \alpha \quad \text{and} \quad \sin 90^\circ = 1$$

Definition IV.9.2 (cosine function for acute and obtuse angles). The ratio $\frac{b}{c}$ of the leg b adjacent the acute angle α to the hypothenuse of a right triangle is the *cosine function* of this angle:

$$\cos \alpha = \frac{b}{c}$$

For right and acute angles, one defines

$$\cos(180^\circ - \alpha) = -\cos \alpha \quad \text{and} \quad \cos 90^\circ = 0$$

Problem IV.9.1. *Convince yourself that*

$$\cos \alpha = \sin(90^\circ - \alpha) \tag{IV.9.1}$$

$$\tan \alpha = \frac{\sin \alpha}{\cos \alpha} \tag{IV.9.2}$$

$$\frac{1}{\tan \alpha} = \tan(90^\circ - \alpha) \tag{IV.9.3}$$

$$\tan(180^\circ - \alpha) = -\tan \alpha \tag{IV.9.4}$$

Problem IV.9.2. *Convince yourself that*

$$\sin^2 \alpha + \cos^2 \alpha = 1 \tag{IV.9.5}$$

is equivalent to the Theorem of Pythagoras

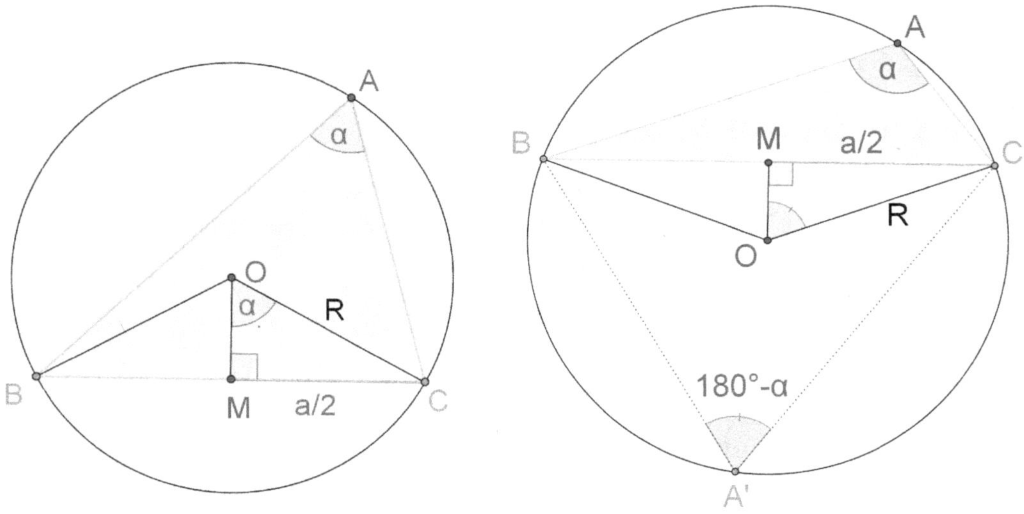

Figure IV.9.1: Proof of the sin theorem for acute, and for obtuse angles.

Theorem IV.9.1 (The extended sin Theorem). *For any triangle with sides a, b, c and angles α, β, γ, and radius R of the circum-circle*

$$\frac{a}{\sin\alpha} = \frac{b}{\sin\beta} = \frac{c}{\sin\gamma} = 2R$$

Proof. This is an easy consequence of (Euclid III.20) and (Euclid III.21). As stated in theorem I.1.4, the central angle is twice the circumference angle.

We put the circum circle around the given triangle. Let its radius be R. We connect the endpoints of side $a = BC$ with the center O, and drop the perpendicular from O this side. Let M be the foot point. The central angle is $\angle BOC = 2\alpha$, since $\angle BAC = \alpha$. It is bisected by the ray $\overrightarrow{OM}$. The side a is bisected by point M. In the right triangle $\triangle OMC$, the definition of the sin function yields

$$\sin\alpha = \frac{|CM|}{|OC|} = \frac{a}{2R}$$

Thus we obtain the required sin theorem for the acute angles of the triangle.

For the obtuse angle, we need to use (Euclid III.22), telling that the opposite angles of a circular quadrilateral $\square ABCA'$ add up to two right angles. We can take for A' any point on the arc opposite to arc BAC, and obtain $\angle BA'C = 180^\circ - \alpha$. This time the central angle $\angle BOC = 2(180^\circ - \alpha)$. Now this angle is bisected by the ray $\overrightarrow{OM}$ and we obtain

$$\sin(180^\circ - \alpha) = \frac{|CM|}{|OC|} = \frac{a}{2R}$$

Hence, in order to obtain a sin theorem valid for both acute and obtuse triangles, one defines

$$\sin(180° - \alpha) = \sin \alpha$$

□

IV.9.2 The Law of Cosines

Theorem IV.9.2 (The cosine theorem). *For any triangle with sides a, b, c and angles α, β, γ*

$$c^2 = a^2 + b^2 - 2ab\cos\gamma$$

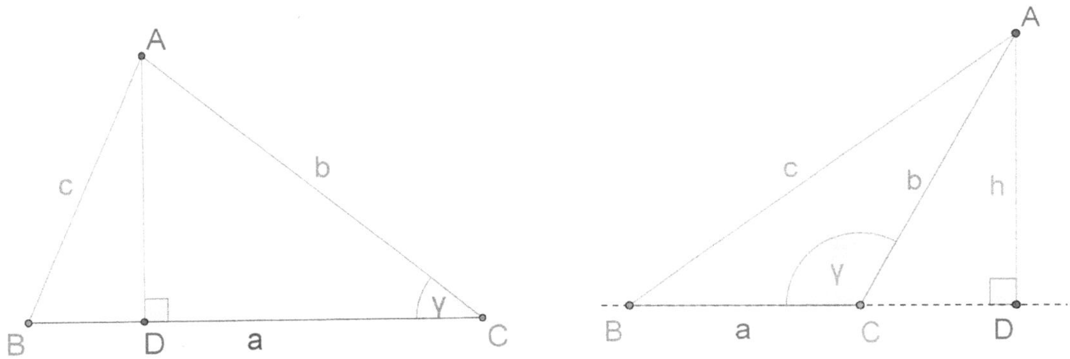

Figure IV.9.2: Proof of the cos theorem for acute, and for obtuse angles.

Proof. We drop the perpendicular from vertex A onto the opposite side a. Let D the foot point. The Pythagorean theorem for triangles $\triangle ADC$ and $\triangle ADB$ yields, after subtracting

$$b^2 = h^2 + |DC|^2$$
$$c^2 = h^2 + |DB|^2$$
$$c^2 = b^2 - |DC|^2 + |DB|^2$$

We need now to distinguish the case of an acute and obtuse triangle. Take an acute triangle. By definition of the cosin function $\cos\gamma = \frac{|DC|}{b}$, and hence $|DC| = b\cos\gamma$.

From the drawing, we see that $|DC| + |DB| = a$. Hence

$$\begin{aligned}
|DB| &= a - |DC| = a - b\cos\gamma \\
|DB|^2 &= a^2 - 2ab\cos\gamma + b^2\cos^2\gamma \\
c^2 &= b^2 - |DC|^2 + |DB|^2 = b^2 - b^2\cos^2\gamma + a^2 - 2ab\cos\gamma + b^2\cos^2\gamma \\
&= a^2 + b^2 - 2ab\cos\gamma
\end{aligned}$$

as required. For the obtuse triangle, we get $|DB| - |DC| = a$. By definition of the cos function $\cos(180° - \gamma) = \frac{|DC|}{b}$, and hence $|DC| = b\cos(180° - \gamma)$. From the drawing, we see that $a = |DB| - |DC|$. Hence

$$\begin{aligned}
|DB| &= a + |DC| = a + b\cos(180° - \gamma) \\
|DB|^2 &= a^2 + 2ab\cos(180° - \gamma) + b^2\cos^2(180° - \gamma) \\
c^2 &= b^2 - |DC|^2 + |DB|^2 \\
&= b^2 - b^2\cos^2(180° - \gamma) + a^2 + 2ab\cos(180° - \gamma) + b^2\cos^2(180° - \gamma) \\
c^2 &= a^2 + b^2 + 2ab\cos(180° - \gamma)
\end{aligned}$$

Hence, in order to obtain a cos theorem valid for both acute and obtuse triangles, one defines

$$\cos(180° - \alpha) = -\cos\alpha$$

□

Corollary 18 (Pythagorean comparison). *For acute angle $\gamma < 90°$, the sides of a triangle satisfy $c^2 < a^2 + b^2$. For obtuse angle $\gamma > 90°$, the sides of a triangle satisfy $c^2 > a^2 + b^2$.*

Corollary 19 (The converse Pythagorean Theorem). *If the sides of a triangle satisfy $c^2 = a^2 + b^2$* in the sense of segment arithmetic, *then the angle across side c is a right angle.*

Problem IV.9.3. *For a triangle are given the sides $a = 10$ and $c = 6$, and the angle $\gamma = 30°$ across the latter side. Find angle α, side b and finally angle β. How many non-congruent solutions do you get?*

Answer. The sin Theorem yields

$$\sin\alpha = \frac{a\sin\gamma}{c} = \frac{5}{6}$$

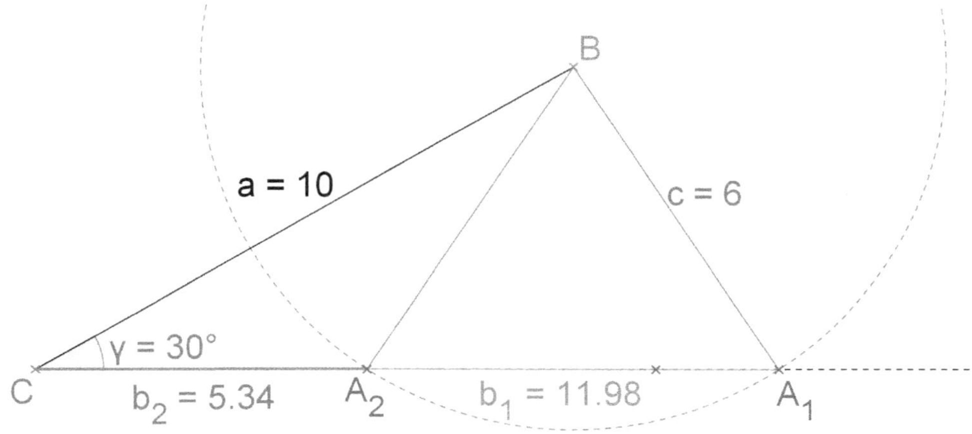

Figure IV.9.3: For the given data, there are two non-congruent solutions.

Hence we get two supplementary angles

$$\alpha_1 = 56.44^\circ \quad \text{and} \quad \alpha_2 = 180^\circ - \alpha_1 = 123.56^\circ$$

as possible solutions. The side b is obtained with the cos Theorem:

$$c^2 = a^2 + b^2 - 2ab\cos\gamma$$
$$b^2 - 10\sqrt{3}b + 64 = 0$$
$$b_{1,2} = \frac{10\sqrt{3} \pm \sqrt{300 - 256}}{2} = 5\sqrt{3} \pm \sqrt{11}$$

Question. It is not yet clear how these results combine, and whether there are two or even four solutions. Provide a drawing to decide this question.

Answer. One sees that the shorter side $b_2 = 5\sqrt{3} - \sqrt{11} = 5.3436$ combines only with the obtuse angle α_2. The longer side $b_1 = 5\sqrt{3} + \sqrt{11} = 11.977$ combines only with the acute angle α_1. Hence one obtains two non-congruent solutions.

Finally, the angle β can be obtained from the angle sum. One gets

$$\beta_1 = 180^\circ - \alpha_1 - \gamma = 93.548^\circ$$
$$\beta_2 = 180^\circ - \alpha_2 - \gamma = 26.45^\circ$$

Question. Obtain the angle β with the sin Theorem. Which problem does arise?

Answer. The sin Theorem yields

$$\sin\beta = \frac{b\sin\gamma}{c} = \frac{5\sqrt{3} \pm \sqrt{11}}{12}$$

One cannot yet know whether the acute or obtuse solution is valid. Indeed $\beta_2 = 26.44$ is acute and $\beta_1 = 93.56$ is obtuse.

Problem IV.9.4. *Use the cosine theorem to give another proof of the parallelogram equation from proposition IV.8.2.*

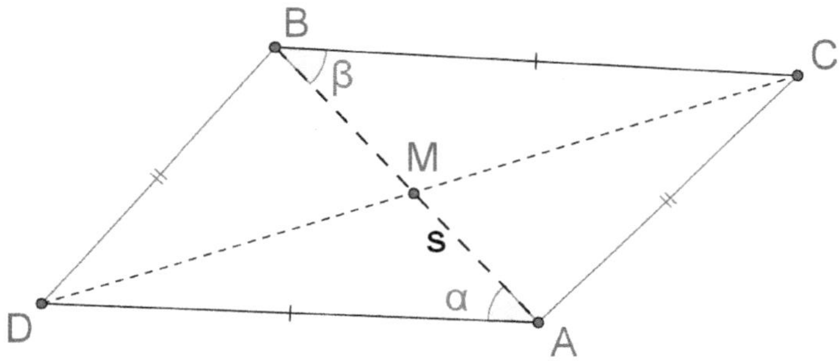

Figure IV.9.4: The diagonals of a parallelogram bisect each other.

Another proof of the parallelogram equation. Refer to the figure on page 179. Given is a parallelogram $\square ACBD$ with two pairs of opposite parallel sides $AC \parallel BD$ and $BC \parallel DA$. Using congruent z-angles and ASA congruence, one sees that the opposite sides are pairwise congruent. Let M be the intersection point of the diagonals. We know already by a problem from neutral geometry that the diagonals of the parallelogram bisect each other.

We apply the cosine theorem to the triangles $\triangle ACM$ and $\triangle DCM$:

$$|AC|^2 = |AM|^2 + |CM|^2 - 2|AM| \cdot |CM| \cdot \cos\theta$$
$$|AD|^2 = |AM|^2 + |DM|^2 - 2|AM| \cdot |DM| \cdot \cos\eta$$

The angle $\theta = \angle AMC$ and $\eta = \angle DMC$ are supplementary and hence $\cos\theta = -\cos\eta$.

Given is a convex quadrilateral with two pairs of congruent opposite sides. We have shown that even in neutral geometry the diagonals bisect each other. Hence holds $|CM| = |DM|$, and adding the relations yields

$$|AC|^2 + |AD|^2 = 2|AM|^2 + 2|CM|^2$$

Similarly, we prove that

$$|BC|^2 + |BD|^2 = 2|BM|^2 + 2|CM|^2$$

Since the diagonals of the parallelogram bisect each other we know that $2|AM|^2 + 2|BM|^2 = |AB|^2$ and $4|CM|^2 = |CD|^2$. Hence adding the two equations yields

$$|AC|^2 + |AD|^2 + |BC|^2 + |BD|^2 = 2|AB|^2 + 2|CD|^2$$

as to be shown. □

Problem IV.9.5. *Get an exact expression for the diameter $2R$ of the circum circle in terms of the three sides a, b, c of a triangle.*

Proof. The cos Theorem yields

$$c^2 = a^2 + b^2 - 2ab\cos\gamma$$
$$\cos\gamma = \frac{a^2 + b^2 - c^2}{2ab}$$

The extended sin theorem gives the diameter of the circum circle

$$\begin{aligned} 2R &= \frac{c}{\sin\gamma} = \frac{c}{\sqrt{1-\cos^2\gamma}} = \frac{2abc}{\sqrt{4a^2b^2 - (2ab\cos\gamma)^2}} \\ &= \frac{2abc}{\sqrt{4a^2b^2 - (a^2+b^2-c^2)^2}} = \frac{2abc}{\sqrt{(2ab + a^2 + b^2 - c^2)}\sqrt{(2ab - a^2 - b^2 + c^2)}} \\ &= \frac{2abc}{\sqrt{((a+b)^2 - c^2)}\sqrt{(c^2 - (a-b)^2)}} \\ &= \frac{2abc}{\sqrt{((a+b+c)(a+b-c)(c-a+b)(c+a-b))}} \end{aligned}$$

□

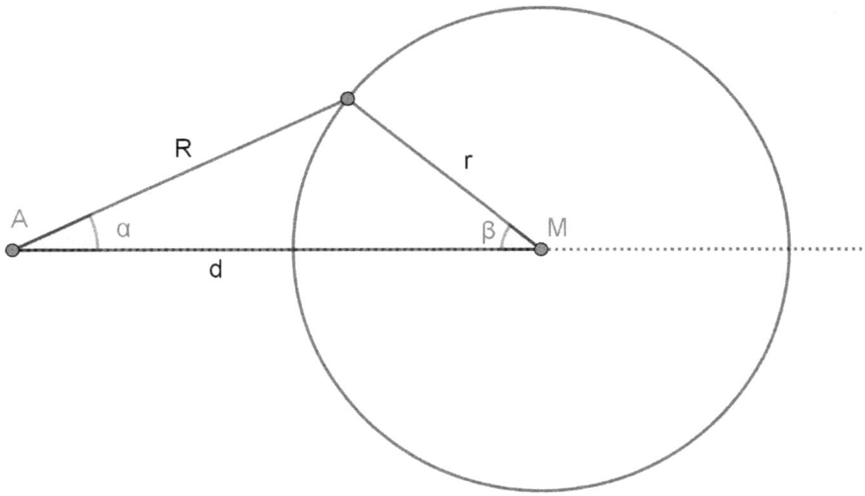

Figure IV.9.5: Attraction of a shell to a point mass.

Problem IV.9.6. *Calculate the attractive gravitational force between a point mass m and a thin spherical shell of radius r and mass M. The point mass has distance d from the center of the shell.*

Solution. In the figure on page 181, the cosine theorem yields

$$R^2 = r^2 + d^2 - 2rd\cos\beta \quad\text{and}\quad r^2 = R^2 + d^2 - 2Rd\cos\alpha$$

Let μ be the mass density per surface angle of the shell, and hence $M = 4\pi\mu$. Integrated over the azimutal angle, the surface element is $2\pi\sin\beta\,d\beta$ and the range is $0 < \beta < \pi$. The attractive force F has only a component along the line connecting the point mass to the center of the shell.

$$F = m\mu\int_0^\pi \frac{\cos\alpha\cdot 2\pi\sin\beta\,d\beta}{R^2}$$

We may eliminate $\cos\alpha$ and get

$$\begin{aligned} F &= \frac{mM}{8\pi d}\int_0^\pi \frac{2Rd\cos\alpha\cdot 2\pi\sin\beta\,d\beta}{R^3} \\ &= \frac{mM}{4d}\int_0^\pi \frac{(R^2+d^2-r^2)\sin\beta\,d\beta}{R^3} \end{aligned}$$

Variable are R or β which depend on each other by $R^2 = r^2 + d^2 - 2rd\cos\beta$ and differentiating gives

$$2RdR = 2rd\sin\beta d\beta$$

The integral for the force is calculated with independent variable R. The range for the new variable is $|d - r| \le R \le d + r$. In the case that $d > r$, i.e. the point mass *outside the sphere*, one gets

$$\begin{aligned} F &= \frac{mM}{4rd^2}\int_0^\pi \frac{(R^2 + d^2 - r^2)rd\sin\beta\, d\beta}{R^3} = \frac{mM}{4rd^2}\int_{d-r}^{d+r} \frac{(R^2 + d^2 - r^2)dR}{R^2} \\ &= \frac{mM}{4rd^2}\left[R - \frac{d^2 - r^2}{R}\right]_{d-r}^{d+r} = \frac{mM}{4rd^2}\left[2r - \frac{d^2 - r^2}{d + r} + \frac{d^2 - r^2}{d - r}\right] \\ &= \frac{mM}{4rd^2}\left[2r - (d - r) + d + r\right] = \frac{mM}{d^2} \end{aligned}$$

as Newton found after a painful long time. □

Problem IV.9.7. *Explain why for the case of a point mass* inside *the spherical shell, there is no attractive force.*

Solution. In this case holds $d < r$ and the range of integration is $r - d \le R \le d + r$. Hence

$$\begin{aligned} F &= \frac{mM}{4rd^2}\left[R - \frac{d^2 - r^2}{R}\right]_{r-d}^{d+r} = \frac{mM}{4rd^2}\left[2d - \frac{d^2 - r^2}{d + r} + \frac{d^2 - r^2}{r - d}\right] \\ &= \frac{mM}{4rd^2}\left[2d - (d - r) - (d + r)\right] = 0 \end{aligned}$$

as Newton found after a painful long time. □

IV.9.3 The Addition Theorem for Tangent

Problem IV.9.8 (The tangent addition Theorem). *Prove the tangent addition theorem*

$$\tan(x + y) = \frac{\tan x + \tan y}{1 - \tan x \, \tan y} \qquad \text{(IV.9.6)}$$

for any acute or obtuse angles x, y with acute or obtuse sum $x + y$.

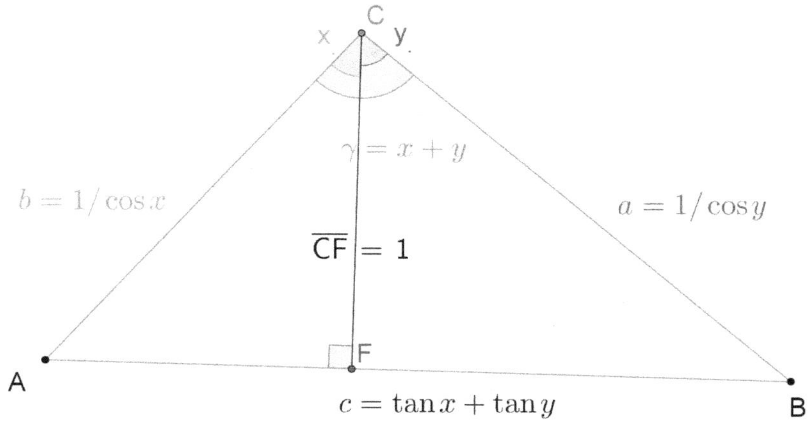

Figure IV.9.6: Proof of the tangent addition theorem.

Proof. We prove the addition theorem at first for any acute or obtuse angles x, y with acute or obtuse sum $\gamma = x + y$. We assume the perpendicular $|FC| = 1$ has unit length and the two right triangles have angles $x = \angle ACF$ and $y = \angle FCB$. From them together, we get the triangle $\triangle ABC$ as shown in the drawing on page 183.

Question. What are the lengths of the sides of the triangle $\triangle ABC$.

Answer.

$$a = |BC| = \frac{1}{\cos y}\ ,\ b = |CA| = \frac{1}{\cos x} \quad \text{and} \quad c = |AB| = \tan x + \tan y$$

Question. Write the cosine theorem for side c in terms of x and y and solve for $\cos(x + y)$.

Answer.

$$\begin{aligned} c^2 &= a^2 + b^2 - 2ab\cos\gamma \\ (\tan x + \tan y)^2 &= \frac{1}{\cos^2 x} + \frac{1}{\cos^2 y} - \frac{2\cos(x+y)}{\cos x \cos y} \\ \cos(x+y) &= \frac{\cos x \cos y}{2}\left[\frac{1}{\cos^2 x} + \frac{1}{\cos^2 y} - (\tan x + \tan y)^2\right] \end{aligned}$$

I use the abbreviation $X = \tan x, Y = \tan y$ and $\cos^2 x = 1/(1 + \tan^2 x)$ to eliminate the cosine function.

$$\cos(x+y) = \frac{\cos x \cos y}{2}\left[2 + X^2 + Y^2 - (X+Y)^2\right] = \cos x \cos y\,[1 - XY]$$

$$1 + \tan^2(x+y) = \frac{(1+X^2)(1+Y^2)}{(1-XY)^2}$$

$$\tan^2(x+y) = \frac{(1+X^2)(1+Y^2) - (1-XY)^2}{(1-XY)^2} = \frac{X^2+Y^2+2XY}{(1-XY)^2}$$

$$\tan(x+y) = \pm\frac{X+Y}{1-XY}$$

Question. Which sign has $1-XY$ for acute angle $x+y$, which for obtuse angle $x+y$. What happens for right angle $x+y$.

Answer. $1-XY>0$ for acute angle $x+y$, and $1-XY<0$ for obtuse angle $x+y$. If and only if $x+y$ is a right angle do we get $1-XY=0$ since this is the case where the two right triangles $\triangle ACF \sim \triangle CBF$ are similar.

Question. How do we justify

$$\tan(180^\circ - \alpha) = -\tan\alpha \tag{IV.9.7}$$

Answer. From the sin and cosine theorems we have obtained

$$\sin(180^\circ - \alpha) = \sin\alpha \quad , \quad \cos(180^\circ - \alpha) = -\cos\alpha$$

for both obtuse and acute angles. Because of the definition

$$\tan\alpha = \frac{\sin\alpha}{\cos\alpha}$$

which we now extend to obtuse angles, we get equation (IV.9.7).

Question. Which sign has $\tan(x+y)$ for acute angle $x+y$, which for obtuse angle $x+y$. What happens for right angle $x+y$.

Answer. $\tan(x+y)>0$ for acute angle $x+y$, and $\tan(x+y)<0$ for obtuse angle $x+y$. If $x+y$ is a right angle, we get $\tan(x+y)=\infty$ because of $\sin 90^\circ = 1$ and $\cos 90^\circ = 0$. [11]

From these considerations of the signs, we see that the same formula

$$\tan(x+y) = \frac{\tan x + \tan y}{1 - \tan x\,\tan y} \tag{IV.9.6}$$

holds for any acute or obtuse angles x, y with acute or obtuse sum $x+y$. □

[11] The meromorphic nature of the tangent function makes a definition meaningful.

IV.9.4 Unwound Angles

Unwound angles occur in the study of the trigonometric functions, physical oscillations, the theory of hyperbolic area and other contexts. The unwound angles have the structure of an ordered Abelian group. This group actually depends on the Hilbert plane considered.

For a precise definition let $\mathcal{A}cute$ denote the set of the congruence classes for the zero and the acute angles. Let R denote the congruence class of right angles.

Definition IV.9.3. The *group of unwound angles* is the Cartesian product $\mathbf{Z}\times\mathcal{A}cute$ with the operation of addition defined by setting

$$(m,\alpha)+(n,\beta)=\begin{cases}(m+n,\alpha+\beta) & \text{if } \alpha+\beta<R\\ (m+n+1,\alpha+\beta-R) & \text{if } \alpha+\beta\geq R\end{cases} \tag{IV.9.8}$$

Proposition IV.9.1. *In any Hilbert plane, the unwound angles have the structure of an ordered Abelian group. This group actually depends on the Hilbert plane considered.*

There is a natural homomorphism from the group of unwound angles onto the group of rotations around the origin. The kernel of this homomorphism is

$$4\mathbf{Z}\times\{0\}=\{(4n,0)\,:\,n\in\mathbf{Z}\}$$

in other words the angles which are integer multiples of $360°$*.*

IV.9.5 Extension to Unwound Angles

Up to now we have defined the sin, cosine and tangent functions for acute and obtuse angles, including the degenerate cases of $0°$ and $180°$.

For the tangent function, we need to allow the infinite value $\tan 90°=\infty$. By allowing ∞ as a value, we extend the tangent function to all unwound angles.

Problem IV.9.9. *Convince yourself that*

$$\tan(90°+x)=-\frac{1}{\tan x} \tag{IV.9.9}$$

holds for $0°\leq x\leq 90°$*.*

Answer. The addition theorem (IV.9.6) for the tangent function implies

$$\tan(45°+x)=\frac{1+\tan x}{1-\tan x}$$

holds for $0° \leq x \leq 135°$. Iterating the relation implies

$$\tan(90° + x) = \frac{1 + \tan(45° + x)}{1 - \tan(45° + x)} = \frac{1 - \tan x + (1 + \tan x)}{1 - \tan x - (1 + \tan x)} = -\frac{1}{\tan x}$$

□

The tangent function is defined for all x by requiring the relation (IV.9.9) for all x.

Proposition IV.9.2. *The sin and cosine functions can be extended to all unwound angles by setting*

$$\cos(180° + x) = -\cos x\,, \quad \sin(180° + x) = -\sin x$$

In that way, $\cos x$ *becomes an even function with period* $360°$, *and* $\sin x$ *becomes an odd function with period* $360°$. *By allowing* ∞ *as a value, we extend the tangent function to all unwound angles: In this way,* $\tan x$ *becomes an odd function with period* $180°$.

Lemma IV.9.1. *We use the imaginary unit* i *which satisfies* $i^2 + 1 = 0$.

$$\cos 2x + i \sin 2x = \frac{(1 + i\tan x)^2}{1 + \tan^2 x} \tag{IV.9.10}$$

$$\cos 2x = \frac{2\tan x}{1 + \tan^2 x} \tag{IV.9.11}$$

$$\sin 2x = \frac{1 - \tan^2 x}{1 + \tan^2 x} \tag{IV.9.12}$$

holds for all unwound angles x. *With the obvious interpretation of the fractions with the value* ∞ *for the tangent function, all values are included.*

Reason. Convince yourself that

$$\cos 2x + i \sin 2x = \frac{1 + i\tan 2x}{\sqrt{1 + \tan^2 2x}} \quad \text{for} \quad -90° < 2x < 90° \tag{IV.9.13}$$

can be checked from the definition. But the positive sign of the root enforces the restriction of the domain. We can use the addition theorem of the tangent and conclude that

$$\tan 2x = \frac{2\tan x}{1 - \tan^2 x} \quad \text{for} \quad -180° < 2x < 180° \tag{IV.9.14}$$

Together the equations (IV.9.13) and (IV.9.14) imply

$$\cos 2x + i \sin 2x = \frac{1 + i \tan 2x}{\sqrt{1 + \tan^2 2x}} = \frac{1 - \tan^2 x + 2i \tan x}{\sqrt{(1 - \tan^2 x)^2 + 4 \tan^2 x}} = \frac{(1 + i \tan x)^2}{1 + \tan^2 x}$$

for $-90° < 2x < 90°$. But we can extend the domain of validity of this equation since it does not contain the root any longer. From the way the extensions of the functions have been defined we get

$$\begin{aligned}\cos 2(90° + x) + i \sin 2(90° + x) &= -\cos 2x - i \sin 2x \\ = -\frac{(1 + i \tan x)^2}{1 + \tan^2 x} = \frac{(\tan x - i)^2}{\tan^2 x + 1} &= \frac{(1 - i/\tan x)^2}{1 + 1/\tan^2 x} = \frac{(1 + i \tan(90° + x))^2}{1 + \tan^2(90° + x)}\end{aligned}$$

Hence equation (IV.9.10) still is valid for $-90° < 2x < 180°$. The formula is extended to $-180° < 2x < 180°$ by the symmetry, and then to all x by the periodicity properties of the trigonometric functions. □

Proposition IV.9.3. *The extended cosine and sin function satisfy the addition theorem*

$$\cos(x + y) + i \sin(x + y) = (\cos x + i \sin x)(\cos y + i \sin y) \tag{IV.9.15}$$

$$\cos(x + y) = \cos x \cos y - \sin x \sin y \tag{IV.9.16}$$

$$\sin(x + y) = \cos x \sin y + \sin x \cos y \tag{IV.9.17}$$

The extended tangent function satisfied the addition theorem

$$\tan(x + y) = \frac{\tan x + \tan y}{1 - \tan x \, \tan y} \tag{IV.9.6}$$

Reason. The addition theorem for the tangent function is extended via formula (IV.9.9). From the original formula valid in the domain $0 \le x, y, x + y \le 180°$ one gets

$$\tan(90° + x + y) = -\frac{1}{\tan(x + y)} = \frac{\tan x \, \tan y - 1}{\tan x + \tan y}$$

$$\frac{\tan(90° + x) + \tan y}{1 - \tan(90° + x) \, \tan y} \frac{\tan x}{\tan x} = \frac{-1 + \tan x \tan y}{\tan x + \tan y}$$

which are equal. With a similar step for y, one extends the domain of validity of formula (IV.9.6) to the domain $0 \le x, y \le 180°$. Now periodicity implies that the formula holds for all x, y.

The addition theorems for sin and cosine follow now via formula (IV.9.10):

$$\begin{aligned}
\cos 2(x+y) + i\sin 2(x+y) &= \frac{[1+i\tan(x+y)]^2}{1+\tan^2(x+y)} \\
&= \frac{[1-\tan x\tan y + i(\tan x+\tan y)]^2}{(1-\tan x\tan y)^2+\tan^2(x+y)} \\
&= \frac{[(1+i\tan x)(1+i\tan y)]^2}{(1+\tan^2 x)(1+\tan^2 y)} = \frac{(1+i\tan x)^2}{1+\tan^2 x}\cdot\frac{(1+i\tan y)^2}{1+\tan^2 y} \\
&= (\cos x + i\sin x)(\cos y + i\sin y)
\end{aligned}$$

□

IV.9.6 Mollweid's Formulas

Given any triangle $\triangle ABC$, the radius of the circum-circle is denoted by R and the radius of the in-circle by ρ. The radii of the ex-circles touching sides a,b or c, respectively, are denoted by ρ_a, ρ_b and ρ_c.

Proposition IV.9.4 (The sin theorem and the in-circle). *For any triangle with sides a, b, c and angles α, β, γ across them*

$$\text{(IV.9.18)} \qquad \begin{aligned}
2R &= \frac{a}{\sin\alpha} = \frac{b}{\sin\beta} = \frac{c}{\sin\gamma} \\
2R &= \frac{\rho}{2\sin\frac{\alpha}{2}\sin\frac{\beta}{2}\sin\frac{\gamma}{2}} \\
&= \frac{\rho_a}{2\sin\frac{\alpha}{2}\cos\frac{\beta}{2}\cos\frac{\gamma}{2}} = \frac{\rho_b}{2\cos\frac{\alpha}{2}\sin\frac{\beta}{2}\cos\frac{\gamma}{2}} = \frac{\rho_c}{2\cos\frac{\alpha}{2}\cos\frac{\beta}{2}\sin\frac{\gamma}{2}}
\end{aligned}$$

Proof. As in the proof of proposition IV.7.2, one gets

$$\cot\frac{\alpha}{2} = \frac{|AF|}{|FI|} \quad\text{and}\quad \cot\frac{\beta}{2} = \frac{|FB|}{|FI|} \quad\text{and hence}\quad \frac{c}{\rho} = \frac{|AB|}{|FI|} = \cot\frac{\alpha}{2} + \cot\frac{\beta}{2}$$

Now one uses the addition theorem and $\alpha+\beta+\gamma = 180°$ to get

$$\frac{\rho}{c} = \frac{\sin\frac{\alpha}{2}\sin\frac{\beta}{2}}{\cos\frac{\alpha}{2}\sin\frac{\beta}{2}+\sin\frac{\alpha}{2}\cos\frac{\beta}{2}} = \frac{\sin\frac{\alpha}{2}\sin\frac{\beta}{2}}{\sin\frac{\alpha+\beta}{2}} = \frac{2\sin\frac{\alpha}{2}\sin\frac{\beta}{2}\sin\frac{\gamma}{2}}{\sin\gamma}$$

To deal with the ex-circles, one derives at first the formula

$$\frac{c}{\rho_a} = \cot\frac{\alpha}{2} - \tan\frac{\beta}{2}$$

and proceeds quite similarly. □

Theorem IV.9.3 (Mollweid's formulas and the tangent theorem). *Sides and angles of any triangle satisfy*

$$\frac{a-b}{c} = \frac{\sin(\alpha-\beta)/2}{\sin(\alpha+\beta)/2} \tag{IV.9.19}$$

$$\frac{a+b}{c} = \frac{\cos(\alpha-\beta)/2}{\cos(\alpha+\beta)/2} \tag{IV.9.20}$$

$$\frac{a+b}{a-b} = \frac{\tan(\alpha+\beta)/2}{\tan(\alpha-\beta)/2} \tag{IV.9.21}$$

Reason for formula (IV.9.19). Use several addition theorems, the angle sum and the cosine theorem three times:

$$\begin{aligned}\frac{\sin(\alpha-\beta)/2}{\sin(\alpha+\beta)/2} &= \frac{2\,[\sin(\alpha-\beta)/2]\cdot[\sin(\alpha+\beta)/2]}{2\sin^2(\alpha+\beta)/2} = \frac{\cos\beta-\cos\alpha}{2\cos^2(\gamma/2)} = \frac{\cos\beta-\cos\alpha}{1+\cos\gamma}\\ &= \frac{b(2ac\cos\beta)-a(2bc\cos\alpha)}{2abc+c(2ab\cos\gamma)} = \frac{b(a^2+c^2-b^2)-a(b^2+c^2-a^2)}{2abc+c(a^2+b^2-c^2)}\\ &= \frac{(b-a)(c^2-(a+b)^2)}{c((a+b)^2-c^2)} = \frac{a-b}{c}\end{aligned}$$

□

Reason for formula (IV.9.20). Use the addition theorem, the angle sum, the sin theorem twice and then the cosine theorem:

$$\begin{aligned}\frac{\sin(\alpha-\beta)}{\sin(\alpha+\beta)} &= \frac{\sin\alpha\cos\beta-\cos\alpha\sin\beta}{\sin\gamma} = \frac{a\cos\beta-b\cos\alpha}{c} = \frac{2ac\cos\beta-2bc\cos\alpha}{2c^2}\\ &= \frac{a^2+c^2-b^2-b^2+c^2+a^2}{2c^2} = \frac{a^2-b^2}{c^2}\end{aligned}$$

Now use the half-angle formula and divide by formula (IV.9.19):

$$\begin{aligned}\frac{2\,[\sin(\alpha-\beta)/2]\cdot[\cos(\alpha-\beta)/2]}{2\,[\sin(\alpha+\beta)/2]\cdot[\cos(\alpha+\beta)/2]} &= \frac{(a-b)(a+b)}{c^2}\\ \frac{\cos(\alpha-\beta)/2}{\cos(\alpha+\beta)/2} &= \frac{a+b}{c}\end{aligned}$$

□

Formula (IV.9.21). —results from dividing formula (IV.9.19) by formula (IV.9.20) □

IV.10 Measurement of the Circle

In this section we shall need the following axioms.

V. Axioms of continuity

V.1 (Axiom of Archimedes) If AB and CD are any segments, then there exists a natural number n such that n segments congruent to CD constructed contiguously from A, along a ray from A through B, will pass beyond B.

V.2 (Cantor's principle of boxed intervals) Every sequence of boxed intervals contains a common point. In detail:

For a sequence of segments A_iB_i such that

$$A_i * A_{i+1} * B_{i+1} * B_i \quad \text{for all} \quad i = 1, 2, 3, \ldots$$

there exists a point X^* such that

$$A_i * X^* * B_i \quad \text{for all} \quad i = 1, 2, 3, \ldots$$

IV.10.1 Approximate Measurement of a Circular Arc

In the interior of the angle $\theta = \angle AOB$, we draw a circular arc of radius $r = 1$. We want to get lower and upper bounds for the length of this arc. Indeed, we want to construct a *convergent sequence* of more and more precise lower and upper bounds.

As a first lower bound, one can simply use the length $a_1 = |AB|$ of the chord. To get an upper bound, we draw the tangents to the circle at the endpoints A and B of the arc. They intersect at point C. The length of the union of the segment AC and CB give the first upper bound $b_1 = |AC| + |CB|$.

Question. Why is $\overrightarrow{OC}$ the bisector of angle $\angle AOB$.

Question. Get the formulas for a_1 and b_1. From which right triangles are they obtained.

Answer. Since $|OA| = |OB| = 1$, we get from right triangles $\triangle OMA$ and $\triangle OAC$

$$a_1 = 2\sin\frac{\theta}{2} \quad \text{and} \quad b_1 = 2\tan\frac{\theta}{2}$$

To get the second, and better lower and upper bounds, we draw the bisector OC of the given angle $\angle AOB$, and mark the point D where it intersects the circular arc

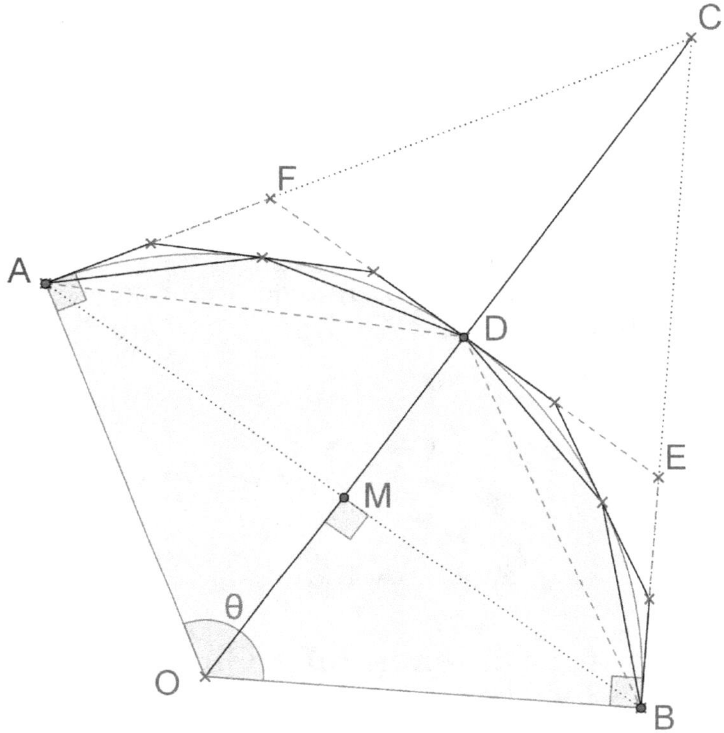

Figure IV.10.1: Archimedes' approximation of a circular arc

$\overset{\frown}{AB}$. The construction of the first lower and upper bounds is now repeated for the two arcs $\overset{\frown}{AD}$ and $\overset{\frown}{DB}$. Since we have doubled the polygons for half the angle, we obtain

$$a_2 = 4\sin\frac{\theta}{4} \quad \text{and} \quad b_2 = 4\tan\frac{\theta}{4}$$

Proposition IV.10.1. *For a short arc θ of the unit circle, the lengths a_n of the inscribed polygon with n sides, and the lengths the b_n of the circumscribed polygon with 2^n sides are calculated recursively: The initial values are*

$$a_1 = 2\sin\frac{\theta}{2} \quad \textit{and} \quad b_1 = 2\tan\frac{\theta}{2}$$

and the two sequences a_n and b_n with $n = 1, 2, \dots$ satisfy the recursion

$$a_{n+1} := 2^{n+1}\sqrt{\frac{b_n - a_n}{2\,b_n}} \quad , \quad b_{n+1} := 2^{n+1}\sqrt{\frac{b_n - a_n}{b_n + a_n}} \tag{IV.10.1}$$

Proof.

Question. Give the trigonometric formulas for a_n and b_n.

$$a_n = 2^n \sin\frac{\theta}{2^n} \quad \text{and} \quad b_n = 2^n \tan\frac{\theta}{2^n} \tag{IV.10.2}$$

To check the recursion, we use the trigonometric formulas

$$\cos\frac{\alpha}{2} = \sqrt{\frac{1+\cos\alpha}{2}} \quad , \quad \sin\frac{\alpha}{2} = \sqrt{\frac{1-\cos\alpha}{2}} \quad , \quad \tan\frac{\alpha}{2} = \sqrt{\frac{1-\cos\alpha}{1+\cos\alpha}}$$

which hold for $\alpha \in [0, \pi]$ with positive roots. □

Problem IV.10.1. *Convince yourself that lengths of the polygons which approximate the circular arc from inside and outside satisfy*

$$a_1 < a_2 < \cdots < a_n < a_{n+1} < \cdots < b_{n+1} < b_n < \cdots < b_2 < b_1 \tag{IV.10.3}$$

Proof.

Question. In which triangles does the triangle inequality yield $a_1 < b_1$, $a_1 < a_2$ and $b_2 < b_1$.

Answer. From the triangle inequality in triangle $\triangle ABC$, we obtain $a_1 < b_1$. From the triangle inequality in triangle $\triangle ABD$, we obtain $a_1 < a_2$. From the triangle inequality in triangle $\triangle EFC$, we obtain $b_2/2 = |EF| < |EC| + |CF| = b_1 - (b_2/2)$ and hence $b_2 < b_1$.

Finally, we get $a_2 < b_2$ and hence together $a_1 < a_2 < b_2 < b_1$. Repeated bisections yields two sequences of polygons which approximate the circular arc from inside and outside. Their lengths satisfy

$$a_1 < a_2 < \cdots < a_n < a_{n+1} < \cdots < b_{n+1} < b_n < \cdots < b_2 < b_1$$

□

For the angle $\theta = 60°$ one gets the initial values

$$a_1 := 1 \quad \text{and} \quad b_1 := \frac{2\sqrt{3}}{3}$$

for the sequences with limit

$$\lim_{n\to\infty} a_n = \lim_{n\to\infty} b_n = \frac{\pi}{3}$$

For a calculation of π all a_n and b_n need to be multiplied with three.

Problem IV.10.2.

$$A_1 := 3 \quad \text{and} \quad B_1 := 2\sqrt{3}$$

$$A_{n+1} := 3 \cdot 2^{n+1} \sqrt{\frac{B_n - A_n}{2B_n}} \ , \quad B_{n+1} := 3 \cdot 2^{n+1} \sqrt{\frac{B_n - A_n}{A_n + B_n}}$$

Compute numerically and report the values of A_n and B_n for $n = 1, 2, \ldots, 15$. Find a conjecture about the trend for the errors $B_n - \pi$ and $\pi - A_n$.

Which value of n gives the best result for π. In which interval can you say π lies for sure?

Answer. As a trend for the errors, one sees $\pi - A_n \cong (B_n - \pi)/2$ and $B_{n+1} - \pi \cong (\pi - A_n)/2$. The trend breaks down for $n = 11$, where the rounding errors makes it impossible to further improve the result. The result for $n = 10$ is still reliable. Multiplication with $3 \cdot 2^{10} = 3\,072$ makes the last three digits random—but one still gets that $3.141592 < \pi < 3.141594$.

n $3 \cdot 2^n$	B_n A_n	$B_n - \pi$ $\pi - A_n$
1 6	3.464101615 3	.3225089615 .1415926536
2 12	3.215390309 3.105828541	.0737976556 .0357641124
3 24	3.159659942 3.132628613	.0180672885 .0089640403
4 48	3.146086215 3.139350203	.0044935615 .0022424506
5 96	3.1427146 3.141031951	.0011219461 .0005607026872
6 192	3.14187305 3.141452472	.0002803964226 .0001401812718
7 384	3.141662747 3.141557608	.0000700935935 .0000350455515
8 768	3.141610177 3.141583892	.000017523088 .0000087613684
9 1 536	3.141597036 3.141590465	.0000043824205 .0000021886731
10 3 072	3.141593778 3.141592135	.0000011239655 .0000005188065
11 6 144	3.141592836 3.141592425	.0000001824518 .000000228241
12 12 288	3.141592234 3.141592131	−.0000004197604 .0000005224336
13 24 576	3.141592458 3.141592432	−.0000001956592 .0000002213274
14 49 152	3.141586207 3.141586201	−.000006446526 .0000064529797
15 98 304	3.141601567 3.141601565	.000008913278 −.0000089116739

Lemma IV.10.1. *The difference of the upper and lower bounds is estimated from above*

$$b_n - a_n < \frac{a_n}{4^{n-1}\, a_1}\,(b_1 - a_1) < \frac{4b_1(b_1 - a_1)}{4^n\, a_1} \tag{IV.10.4}$$

Problem IV.10.3. *Explain how to get the estimate* (IV.10.4).

Reason.

Question. Check the following formula:

$$a_n^{-2} - b_n^{-2} = 4^{-n}$$

Question. We define the quotients $q_n := \frac{b_n}{a_n}$. Use the trigonometric formula $\cos^2 \frac{\alpha}{2} = \frac{1+\cos\alpha}{2}$ to check the recursion formula

$$q_{n+1}^2 = \frac{2q_n}{1+q_n}$$

Question. Prove that $0 \le q_{n+1} - 1 < (q_n - 1)/4$ for all $n \ge 1$.

Answer. Since $0 < a_n < b_n$ division implies $q_n > 1$ for all $n \ge 1$.

$$q_{n+1}^2 - 1 = \frac{2q_n}{q_n + 1} - 1 = \frac{q_n - 1}{q_n + 1}$$

$$q_{n+1} - 1 = \frac{q_n - 1}{(q_n + 1)(q_{n+1} + 1)} < \frac{q_n - 1}{4}$$

The estimate of q_n implies

$$b_{n+1} - a_{n+1} = a_{n+1}(q_{n+1} - 1) < \frac{a_{n+1}}{4}(q_n - 1) = \frac{a_{n+1}}{4\, a_n}(b_n - a_n)$$

For all $n \ge 1$. Finally one gets the result by induction. □

Problem IV.10.4. *Use the estimate* (IV.10.4) *and explain which continuity axioms are needed to conclude that both sequences b_n and a_n converge to the same limit.*

Answer. As far as the continuity axioms are concerned, one needs to assume the Archimedean axiom (V.1) IV.10 and Cantor's principle of boxed intervals (V.2) IV.10. See the section on measurement and continuity.

Because of the estimates (IV.10.3), Cantor's principle of boxed intervals implies there exists a length x^* such that

$$a_1 < a_2 < \cdots < a_n < x^* < b_n < \cdots < b_2 < b_1 \quad \text{for all natural } n. \tag{IV.10.5}$$

Moreover, the estimate (IV.10.4) together with the Archimedean axiom (V.1) imply that the number x^* is unique. Indeed, assume that both x^* and x satisfy the estimate (IV.10.5). Hence

$$|x^* - x| \le b_n - a_n < \frac{4b_1(b_1 - a_1)}{4^n\, a_1} = K \cdot 4^{-n}$$

hold for all natural n. This is only possible if $x = x^*$.

Reason. One proves by induction that $N < 2^N \leq 4^N$ holds for all natural N. Assume towards a contradiction that $x^* \neq x$. By the Archimedean axiom there would exist a natural number N such that $K < N|x^* - x|$. Hence $|x^* - x| < N|x^* - x|4^{-n}$ and $4^n < N$ would hold for all natural n. This is false even just for $n = N$. □

Remark. The axioms of continuity used above are my own favorites since they can be used directly in the proofs. Moreover they are the axioms of continuity contained in the very first edition of Hilbert's Foundations of Geometry. [12]

IV.10.2 Arc Length and Area of a Circular Sector

Definition IV.10.1 (Pi). The arc length of a semicircle of radius one is defined to be the number π.

Definition IV.10.2 (Measurement of angle by arc length). The arc length of a circular arc of radius one and central angle θ is defined as the *arc measurement* of the central angle.

Proposition IV.10.2. *Assume that the Archimedean Axiom holds. The measurement of angles by degrees and arc length are proportional. They are related by*

$$\theta[arc] = \frac{\pi}{180^\circ}\,\theta[^\circ] \quad \textit{and} \quad \theta[^\circ] = \frac{180^\circ}{\pi}\,\theta[arc] \tag{IV.10.6}$$

Measurement of angle by arc length. The *arc length* of the circular arc with radius one and central angle $\theta = \angle AOB$ is obtained from the common limit of the sequences b_n and a_n

$$\theta[arc] = \lim_{n\to\infty} a_n = \lim_{n\to\infty} b_n$$

Consider at first the special case that the angle $\theta = p2^{-n} \cdot 90^\circ$ with integer p and integer $n \geq 0$. In other words, one begins with a right angle, and uses repeated bisection and takes any integer multiples of the angles one has thus obtained. In all these cases, the simple proportionality given by equation (IV.10.6) follows directly from the definition of the number π.

The general case is obtained using the process for measurement of angles explained in the main theorem about measurement in volume I. Such a measurement process presupposes the Archimedean property for angles, stated in theorem V.12.2. Indeed any angle may be approximated, with arbitrary good accuracy by a sequence of angles $\theta = p2^{-n} \cdot 90^\circ$. Hence the proportionality given by equation (IV.10.6) remains valid for any arbitrary angle. □

---

[12] In later editions of Hilbert's foundations, his (rather convoluted) axiom of completeness is postulated.

Theorem IV.10.1 (Circumference and area of a circular arc). *A circular arc with radius r and central angle θ, to be measured in arc length, has the* arc length $\theta\, r$ *and the* area $\theta\, \frac{r^2}{2}$.

Theorem IV.10.2 (Circumference and area of a circle). *The circle of radius r has the circumference $2\pi r$ and the area πr^2.*

Corollary 20 ("pizza theorem"). *The area of a circle is the product of the radius with half of the circumference.*

Problem IV.10.5. *Explain the nickname "pizza theorem".*

The arc length of a circular sector. The approximate measurement of the arc length of a circular arc of radius r is performed as explained in the previous section. One obtains the arclength from the common limit

$$\lim_{n\to\infty} a_n \cdot r = \lim_{n\to\infty} b_n \cdot r = \theta[arc] \cdot r$$

because the measurement of angles by arc length from definition IV.10.2. □

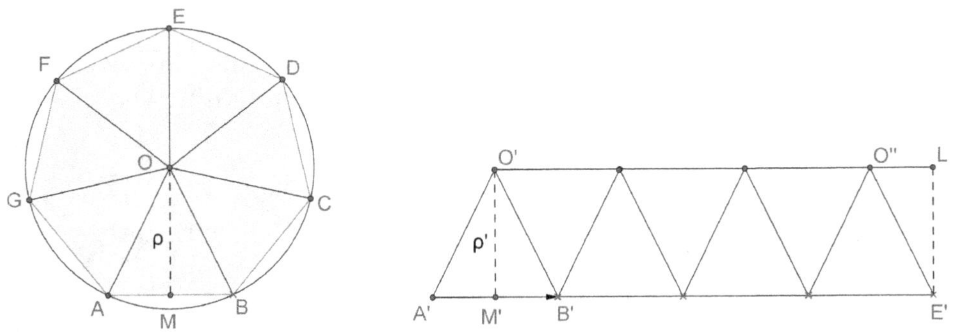

Figure IV.10.2: Measuring the area of a regular polygon.

Lemma IV.10.2. *The area of a regular polygon is the product of the radius ρ of the in-circle with half perimeter s. The area of a regular polygon circumscribed around a circular sector equals the radius times half the length of the polygon.*

Proof. The reasoning is explained for a 7-gon by the figure on page 197. The 7-gon is partitioned into 7 congruent isosceles triangles $\triangle ABO$ through $\triangle GAO$, where point O is the center of the circum-circle. These triangle are rearranged into a trapezoid

$\square A'E'O''O'$. One gets a trapezoid from an n-gon with an odd number n of sides, and one gets a parallelogram from an n-gon with an even number of sides. In both cases, in order to obtain a rectangle, one still needs to bisect the first triangle $\triangle ABO$ into two right triangles, and shift one of them to the opposite side $E'O''$. The rectangle $\square M'E'LO'$ that one has obtained, has the side lengths $|M'O'| = \rho$ which is the radius of the in-circle of the original polygon, and $|M'E'| = s$ which is half of the perimeter of the original polygon. $\square$

The area of a circular sector. To obtain upper- and lower bounds for the area of a circular sector, one uses the areas of the regular polygons circumscribed and inscribed into the circular sector.

By lemma IV.10.2, the area of a regular polygon circumscribed around a circular sector equals the radius r times half the length of the polygon. Hence one gets the area

$$r \cdot \frac{b_n r}{2}$$

as an <u>upper bound</u> of the area of the circular sector.

To obtain a lower bound, we use the area of a regular polygon inscribed into the circular sector. The arc length of the inscribed polygon is $a_n \cdot r$. The radius of the circle inscribed into the inscribed polygon is $\frac{a_n}{b_n} \cdot r$. Hence one gets the area

$$\frac{a_n r}{b_n} \cdot \frac{a_n r}{2}$$

as an <u>lower bound</u> of the area of the circular sector. Using the formulas (IV.10.2) and the identity $\sin \alpha = 2 \sin \frac{\alpha}{2} \cos \frac{\alpha}{2}$, one can simplify

$$\begin{aligned} \frac{a_n r}{b_n} \cdot \frac{a_n r}{2} = \frac{a_n^2}{b_n} \cdot \frac{r^2}{2} &= 2^n \sin \frac{\theta}{2^n} \cos \frac{\theta}{2^n} \cdot \frac{r^2}{2} \\ &= 2^{n-1} \sin \frac{\theta}{2^{n-1}} \cdot \frac{r^2}{2} = \frac{a_{n-1}\, r^2}{2} \end{aligned}$$

Altogether, we have obtained the bounds

$$\frac{a_{n-1}\, r^2}{2} < \text{area of circular sector} < \frac{b_n r^2}{2}$$

and in the limit $n \to \infty$, we obtain

$$\text{area of circular sector} = \lim_{n\to\infty} a_n \cdot \frac{r^2}{2} = \lim_{n\to\infty} b_n \cdot \frac{r^2}{2} = \frac{\theta r^2}{2}$$

with the central angle θ measured in arc length. $\square$

Remark. A particular simple "proof" of the pizza theorem is as follows:

> We know that for a regular n-gon, the area is one-half the product of the radius r of the in-circle to the circumference. "Now image that we make the number of sides of the polygon *infinite*. Each triangle has height r and an *infinitesimal* base dr. Since the polygon fills the circle, their areas are equal. Also, the perimeter of the polygon equals the circumference of the circle. Thus the formula in the pizza theorem follows from the corresponding formula for the regular polygons with any number n of sides."

This idea is usually attributed to Johannes Kepler, but similar arguments were already used by Democritus and other early Greek mathematicians.

Because of the ancient Greeks "horror of the infinite" it should come as no surprise that no important mathematician in ancient times,—even those who devised such proofs,—did considered them to be rigorous. The same can be said of all mathematicians since then, including contemporary ones. Currently, we make the proof acceptable as done above: We give lower and upper estimates for the circumference and area. We use a "squeezing argument" which takes advantage of the fact that the lower and upper estimates have the same limit for $n \to \infty$. In this way the result is obtained from a limit $n \to \infty$ for which the number of sides of the polygon approach infinity. To be even more exact, one has to point out,—as we have done above,—which axioms of continuity are involved to confirm existence and uniqueness of such a limit.

The Greeks did not use these terms. But they had their *method of exhaustion* that was quite similar to the modern rigorous use of limits. This method was first clearly articulated by Eudoxus around 370 B.C. and extensively used by Archimedes.

However, without the modern calculus, the method of exhaustion gives no indications how to discover formulas for areas and volumes in the first place. It just needed the genius of Archimedes for every such result! But the mystery about the way Archimedes worked was (at least partly) resolved when a document by Archimedes called "The Method" was discovered around 1903. This document describes an ingenuous procedure in which one or more solids could be mentally sliced into infinitesimally thin pieces. By achieving a balance piece by piece, and using the basic physical law of levers, Archimedes was able to find the volume of the sphere, and similar results.

This *method of equilibrium* is one of the cleverest integration tricks ever devised. As yet, Archimedes did not even mention,—let alone use,—the method in any of his other works, because he did not consider it to be rigorous.

IV.11 Area in Euclidean Geometry

IV.11.1 Equidecomposable and Equicomplementable Figures

Definition IV.11.1. A *rectilinear figure*, or simply *figure*, is a finite union of non-overlapping triangles. To express that the figure P is the union of non-overlapping triangles T_i for $i = 1 \dots n$, we write

$$P = \biguplus_{i=1}^{n} T_i$$

We call the union $\biguplus T_i$ a *dissection* of the figure P.

Two triangles (or figures) are *non-overlapping* if their intersection is at most a finite union of points and segments.

Definition IV.11.2. Two rectilinear figures P and Q are *equidecomposable* means they are finite unions of non-overlapping, pairwise congruent triangles. Too, we say that the figures P and Q can be *dissected* into congruent triangles, or simply, *have congruent dissections.*

In this case, we write

$$P = \biguplus_{i=1}^{n} T_i \quad \text{and} \quad Q = \biguplus_{i=1}^{n} S_i \quad \text{with} \quad T_i \cong S_i \quad \text{for} \quad i = 1 \dots n$$

Definition IV.11.3. Two rectilinear figures P and Q are *equicomplementable* means there exist figures R and S such that

(1) R and S are equidecomposable

(2) P and R are non-overlapping

(3) Q and S are non-overlapping

(4) $P \uplus R$ and $Q \uplus S$ are equidecomposable

Remark. As a short hand, I write

$P \overset{\boxtimes}{\sim} Q$ for figures P and Q being equidecomposable.

$P \triangleq Q$ to indicate that the figures P and Q are equicomplementable.

Remark. More in everyday language, one may explain as follows: Two figures are equicomplementable means that one can obtain equidecomposable figures from them, by adding to each one an appropriately chosen extra figure. These two extra figures need to be equidecomposable and may not overlap with the two originally given figures.

Remark. In German, Hilbert uses the words "zerlegungsgleich" for equidecomposable, and "ergänzungsgleich" for equicomplementable.

We use some results from the *neutral* theory of area, a more complete treatment of which is given in volume I.

Theorem IV.11.1. *Any two congruent figures are equidecomposable. Any two equidecomposable figures are equicomplementable.*

All four notions—equality, congruence, being equidecomposable as well as being equicomplementable—are equivalence relations among figures. Indeed these are four different *equivalence relations.*

As well known from everyday life, it is more practical to measure the area of figures, instead of only comparing different figures. As a first step in the discussion of area, we state the properties which need to hold for a useful notion of area. Secondly, we give a construction of the area function and thus prove its existence.

Proposition IV.11.1. *If the figures P_i are equidecomposable to the figures Q_i for $i = 1 \ldots n$, and neither the figures P_i nor the figures Q_i overlap, then the union $\biguplus_{i=1}^n P_i$ is equidecomposable to the union $\biguplus_{i=1}^n Q_i$.*

The corresponding statement holds in the case of equicomplementable figures.

Definition IV.11.4. The *area* or *measure* is a function assigning to any rectilinear figure P a number area(P) such that the following properties hold:

(1) area$(T) > 0$ for all triangles T

(2) $T \cong S$ implies area$(T) =$ area(S) for any two triangles

(3) area$(P \uplus Q) =$ area$(P) +$ area(Q) for any non-overlapping figures P and Q.

More generally, the value area(P) can be an element of any ordered Abelian group.

Theorem IV.11.2. (a) *Any two equidecomposable figures have the same area.*

(b) *Any two equicomplementable figures have the same area.*

Of course, in the Euclidean case, the Saccheri and Lambert quadrilaterals are simply rectangles. Given is a triangle $\triangle ABC$, a side AB of which has been chosen as base. I call the line through the midpoints of the two remaining sides a *midline.*

Proposition IV.11.2 (A neutral version of Euclid I.37). *Two triangles with the same base and the same midline are equicomplementable.*

Two triangles with the same base, and the midpoints of the four remaining sides lying on one line, are equicomplementable.

Remark. Astonishing enough, proposition IV.11.2 is even valid in neutral geometry. On the other hand, Euclid's original proposition I.37 is not valid in hyperbolic geometry.

Proof. The proof exploits the correspondence between a triangle and a Saccheri quadrilateral given in Hilbert's Proposition 39. Complete details are given in volume III, the section about Legendre's geometry.

For a given triangle $\triangle ABC$ and base AB, we obtain the corresponding Saccheri quadrilateral $\square FGBA$ by the construction given in the figure on page 202. We have

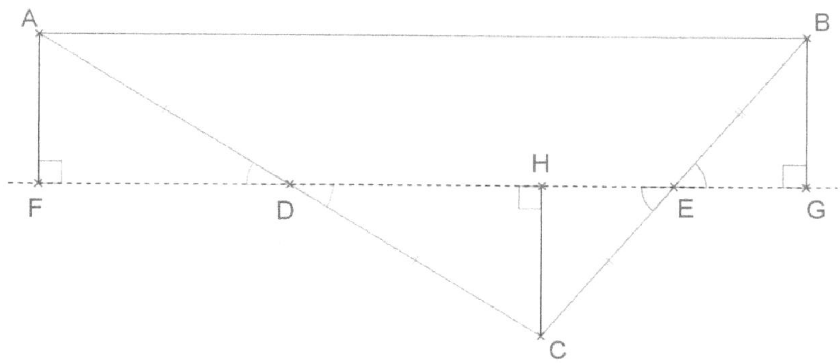

Figure IV.11.1: To every triangle corresponds a Saccheri quadrilateral.

shown by SAA congruence that

$$\triangle AFD \cong \triangle CHD \quad \text{and} \quad \triangle BGE \cong \triangle CHE$$

Problem IV.11.1. *Prove that the segment DE which the triangle cuts out of the middle line has length half the base FG of the corresponding Saccheri quadrilateral.*

Question. Take the case of an acute triangle and convince yourself that the triangle $\triangle ABC$ and the Saccheri quadrilateral $\square FGBA$ are equidecomposable.

We can leave a short answer to this question to the reader. The Saccheri quadrilateral depends only on the base AB and the midline DE of the triangle. Hence we obtain the same Saccheri quadrilateral from another triangle with the same base and same midline. Using the positive answer to the question above, by transitivity (see Theorem IV.11.1) we conclude that two acute triangles with the same base and midline are equidecomposable.

In the case of an obtuse triangle, it can happen that the triangle $\triangle ABC$ and the Saccheri quadrilateral $\square FGBA$ are only equicomplementable. The details are explained below. By transitivity, now used for the relation of equicomplementarity, we conclude that any two triangles with the same base and midline are equicomplementable. □

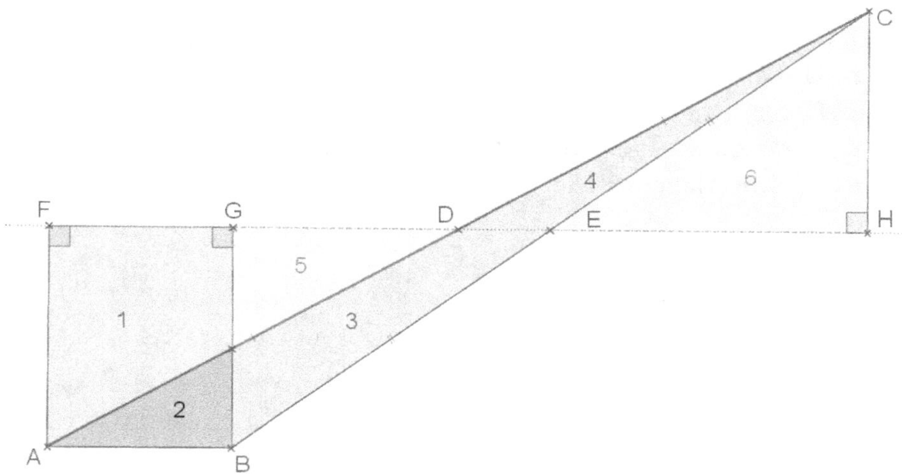

Figure IV.11.2: Even an obtuse triangle is equicomplementable to its corresponding Saccheri quadrilateral.

Question. Take the case of an obtuse triangle, as given in the figure on page 203. Explain why the triangle $\triangle ABC$ and the Saccheri quadrilateral $\square FGBA$ are equicomplementable. Let the numbers denote non-overlapping regions.

Answer. Triangle congruences yield

$$1 \uplus 5 \cong 4 \uplus 6 \quad \text{and} \quad 3 \uplus 5 \cong 6$$

The Saccheri quadrilateral and the triangle

$$\square ABGF \overset{\boxtimes}{\sim} 1 \uplus 2 \quad \text{and} \quad \triangle ABC \overset{\boxtimes}{\sim} 2 \uplus 3 \uplus 4$$

are complemented by the triangular region 5 as additional figure. We conclude

$$\begin{aligned}\square ABGF \uplus 5 &\overset{\boxtimes}{\sim} (1 \uplus 2) \uplus 5 \overset{\boxtimes}{\sim} 2 \uplus (1 \uplus 5) \overset{\boxtimes}{\sim} 2 \uplus (4 \uplus 6) \\ &\overset{\boxtimes}{\sim} (2 \uplus 4) \uplus 6 \overset{\boxtimes}{\sim} (2 \uplus 4) \uplus (3 \uplus 5) \overset{\boxtimes}{\sim} (2 \uplus 4 \uplus 3) \uplus 5 \\ &\overset{\boxtimes}{\sim} \triangle ABC \uplus 5\end{aligned}$$

Hence the Saccheri quadrilateral and the triangle are equicomplementable.

After these remarks about area in neutral geometry, I specialize the Euclidean geometry.

Problem IV.11.2. *Draw two parallelograms on the same base with congruent altitudes, and explain why they are equidecomposable. Clearly draw the dissections you use.*

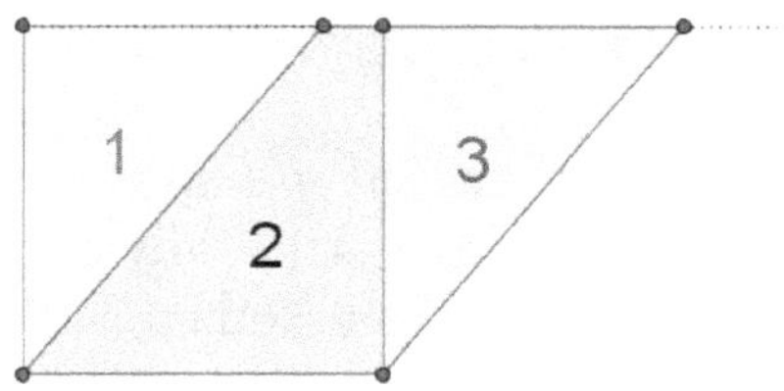

Figure IV.11.3: A square and a parallelogram that are easily seen to be equidecomposable.

Answer. In the figure on page 204, the square is decomposed as $1 \uplus 2$. The parallelogram on the same base is decomposed as $2 \uplus 3$. Since the triangles $1 \cong 3$ are congruent, the square and the parallelogram as equidecomposable.

Problem IV.11.3. *Draw two parallelograms on the same base with congruent altitudes that are not easily seen to be equidecomposable. Explain why they are equicomplementable. Clearly draw the dissections and mark the additional figures used.*

Answer. In the figure on page 205, the square is decomposed as $1 \uplus 2$. The parallelogram on the same base is decomposed as $2 \uplus 3$. But this time 1 and 3 are not congruent.

To see that the square and the parallelogram are equicomplementable, we use the triangle 4 as additional figure. Since the unions

$$1 \uplus 4 \cong 3 \uplus 4$$

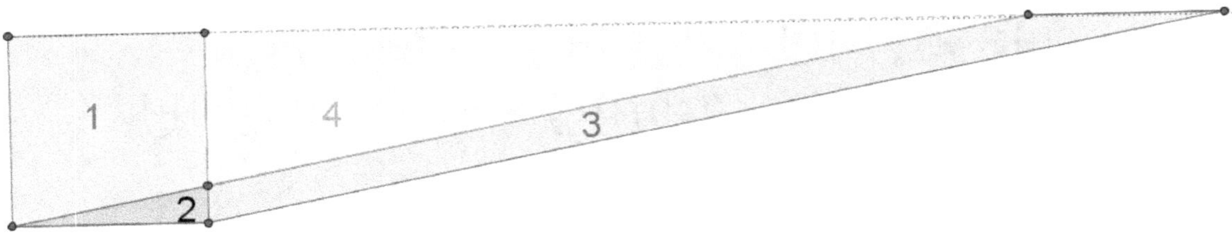

Figure IV.11.4: A square and a parallelogram that are only shown to be equicomplementable.

are indeed congruent, we conclude

$$\begin{aligned} \text{square} \uplus 4 &\overset{\boxtimes}{\sim} (1 \uplus 2) \uplus 4 \overset{\boxtimes}{\sim} 2 \uplus (1 \uplus 4) \overset{\boxtimes}{\sim} 2 \uplus (3 \uplus 4) \\ &\overset{\boxtimes}{\sim} (2 \uplus 3) \uplus 4 \overset{\boxtimes}{\sim} \text{ parallelogram } \uplus 4 \end{aligned}$$

Hence the square and the parallelogram—with the same base and congruent altitudes—are equicomplementable. Similarly, we prove that any two parallelograms with the same base and congruent altitudes are equicomplementable.

We can now give the exact modern version for the propositions about area in Euclid's book one.

Proposition IV.11.3 (Instead of Euclid I.35). *Two parallelograms on the same base with congruent altitudes are equicomplementable.*

Proposition IV.11.4 (Instead of Euclid I.36). *Two parallelograms with congruent bases and congruent altitudes are equicomplementable.*

Problem IV.11.4. *Explain why two parallelograms with the same base and congruent altitudes are always equicomplementable.*

Proposition IV.11.5 (Instead of Euclid I.37). *Two triangles with the same base and congruent altitudes are equicomplementable.*

Remark. Euclid talks about triangles "in the same parallels". Euclid I.37 says literally that `"Triangles on the same base in the same parallels are equal."`

Proposition IV.11.6 (Instead of Euclid I.38). *Two triangles with congruent bases and congruent altitudes are equicomplementable.*

Problem IV.11.5. *Convince yourself by an informal argument that two triangles with the same base and congruent altitudes are equicomplementable.*

Problem IV.11.6. *Use the neutral version of Euclid I.37 given by proposition IV.11.2 above to prove in* <u>Euclidean</u> *geometry that two triangles with the same base and congruent altitudes are equicomplementable.*

Answer. In Euclidean geometry, <u>but only in Euclidean geometry</u>, two triangles with the same base and congruent altitudes have the same midline. (Work out further details!) Hence by proposition IV.11.2 above they are equicomplementable, too.

IV.11.2 The Theorem of Pythagoras and Related Results

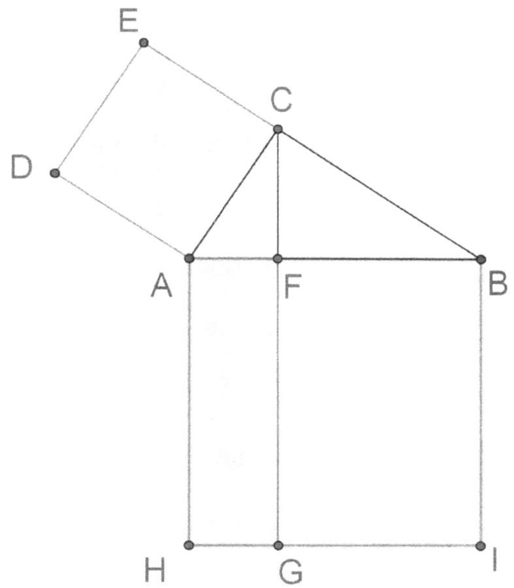

Figure IV.11.5: Euclid's leg theorem tells the square over the leg is equicomplementable to the rectangle as indicated.

As a first step towards the Pythagorean Theorem, Euclid proves the following version of the leg theorem:

Proposition IV.11.7 (Euclid's leg theorem). *The square on a leg of a right triangle is equicomplementable to the rectangle one side of which is the hypothenuse and the other side is the projection of this leg onto the hypothenuse.*

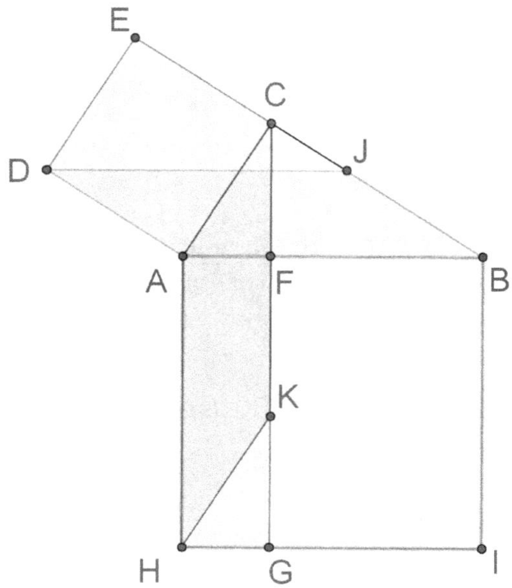

Figure IV.11.6: Euclid's proof of the leg theorem.

Modern proof following Euclid. For a given right triangle $\triangle ABC$, we have constructed the square $\square ACED$ over the leg AC and the square $\square ABIH$ over the hypothenuse AB. We have dropped the altitude from vertex C onto the hypothenuse and got the projection AF of the leg AC. Finally we have obtained the rectangle $\square AFGH$, one side of which is the hypothenuse and the other side is the projection AF of the leg AC onto the hypothenuse AB.

To show that the square $\square ACDE$ and the rectangle $\square AFGH$ are equicomplementable, we use as intermediate steps the two parallelograms $\square ABJD$ and $\square AHKC$. Here point J can be obtained by transferring the leg $AD \cong BJ$, and point K can be obtained by transferring the altitude $FC \cong GK$.

Question. Convince yourself that both quadrilaterals $\square ABJD$ and $\square AHKC$ are indeed parallelograms.

Question. Convince yourself that these two parallelograms are congruent. How can one easily obtain parallelogram $\square AHKC$ from parallelogram $\square ABJD$.

Question. Why is the square $\square ACED$ equicomplementable to the parallelogram $\square ABJD$.

Answer. These two parallelograms have the same base AD and congruent altitudes. Hence they are equicomplementable by Euclid I.37 (see proposition IV.11.5).

Question. Why is the parallelogram $\square ACKH$ equicomplementable to the rectangle $\square AFGH$.

Answer. These two parallelograms have the same base AH and congruent altitudes. Hence they are equicomplementable by Euclid I.37 (see proposition IV.11.5).

By these steps, we have obtained the following chain of equicomplementable figures:

$$\square ACED \triangleq \square ABJD \cong \square ACKH \triangleq \square AFGH$$

By Theorem IV.11.1, any two congruent figures are equicomplementable. Furthermore, the relation "equicomplementable" is an equivalence relation among figures. Hence we conclude that

$$\square ACED \triangleq \square AFGH$$

as to be shown. □

Proposition IV.11.8 (Euclid's (weaker) Pythagorean theorem). *The sum of the squares on the legs of a right triangle is equicomplementable to the square over the hypothenuse.*

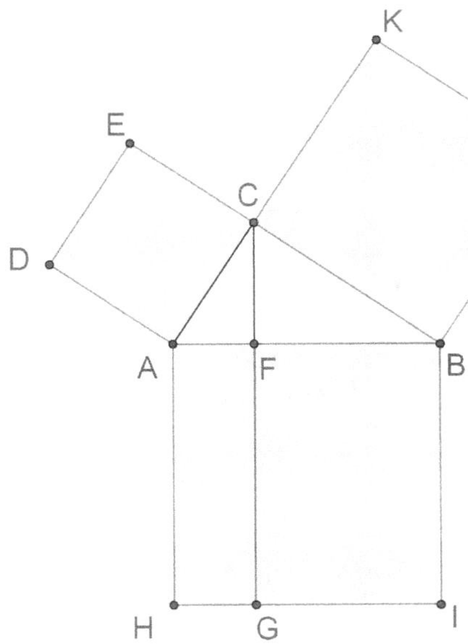

Figure IV.11.7: Euclid's proof of the Pythagorean Theorem.

Modern proof following Euclid. We use Euclid's leg theorem IV.11.7 twice, for both legs:

$$\square ACED \triangleq \square AFGH \quad \text{and} \quad \square BCKJ \triangleq \square BFGI$$

Since these figures do not overlapping we conclude

$$\square ACED \uplus \square BCKJ \triangleq \square AFGH \uplus \square BFGI \triangleq \square ABIH$$

as to be shown. □

Theorem IV.11.3 (The Pythagorean Theorem). *The sum of the squares on the legs of a right triangle is equidecomposable to the square over the hypothenuse.*

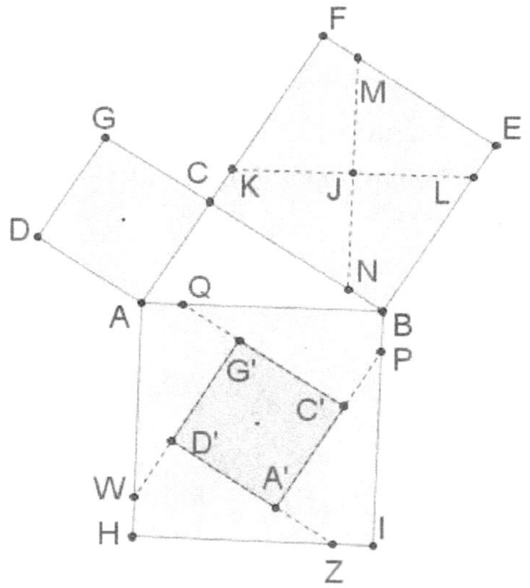

Figure IV.11.8: What went wrong?

Problem IV.11.7. *We want to prove the Theorem of Pythagoras by dissection. What went wrong in the above figure on page 209 to get an appropriate dissection. Draw a figure with a more useful dissection.*

Problem IV.11.8. *Prove the Theorem of Pythagoras using the figure on page 210. For a proof by dissection, we need to*

- *give a construction of this diagram which explains in which order it was obtained.*

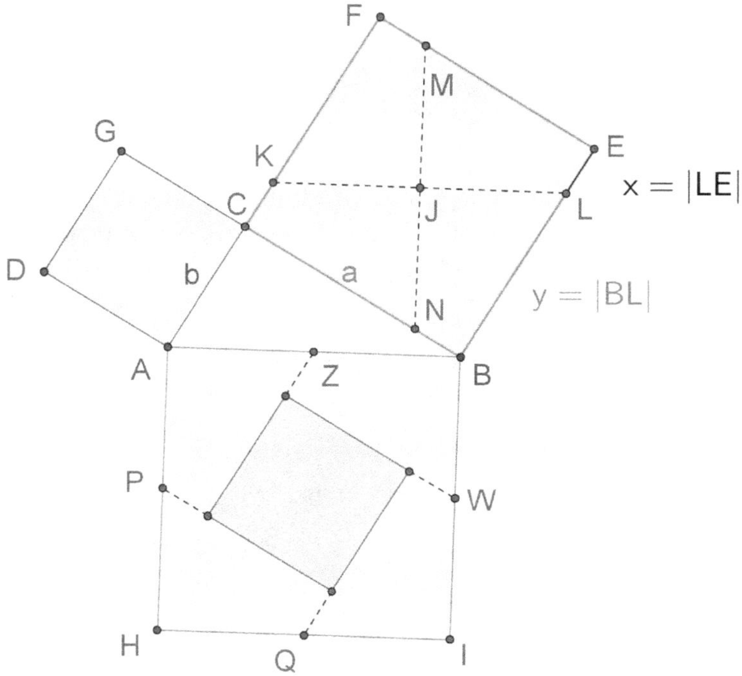

Figure IV.11.9: A dissection proof of the Theorem of Pythagoras.

- *Based on this construction, we need to prove the congruence of the pieces to which the square on the hypothenuse is dissected to the pieces of which the two squares erected on the legs are made up.*

Proof. Here is a way to do these steps: We construct the center J of the square $\square BCFE$ on the larger leg of the given right triangle $\triangle ABC$. Drop the perpendicular from J onto the hypothenuse AB and erect the "double perpendicular" at point J. The square $\square BCFE$ cuts the segments NM and KL out of these two perpendicular lines.

We need to determine the segment lengths $|CK| = x$ and $|BL| = y$. We see that $y + x = |BL| + |LE| = a$. The parallelogram $\square ABLK$ has two pair of opposite congruent sides. Hence $|KL| = |AB| = c$ and $b + x = |AC| + |CK| = |AK| = |BL| = y$.

Since $|KJ| = |JL| = \frac{c}{2}$, the four quadrilaterals into which the square $\square BCFE$ is dissected, are congruent. They are transferred by several different parallel shifts to the square $\square ABIH$ on the hypothenuse. We do this in such a manner that the four right angles at vertex J are transferred to the four vertices A, B, I, H. Hence the transferred quadrilaterals have vertices at the midpoints P, Q, W, Z of the sides of the square $\square ABIH$. Furthermore, they have each one vertex A', C', G' or D' lying inside the square.

We get right angles at these four points. Since $A'C' \cong C'G' \cong G'D' \cong D'A' = y - x = b$, we see that the square $\square A'C'G'D'$ is congruent to the square $\square ACGD$ erected over the shorter leg AC.

Hence the square over the shorter leg, and the four congruent quadrilaterals into which the square over the longer leg has been dissected are put together without overlapping to form the square over the hypothenuse. $\square$

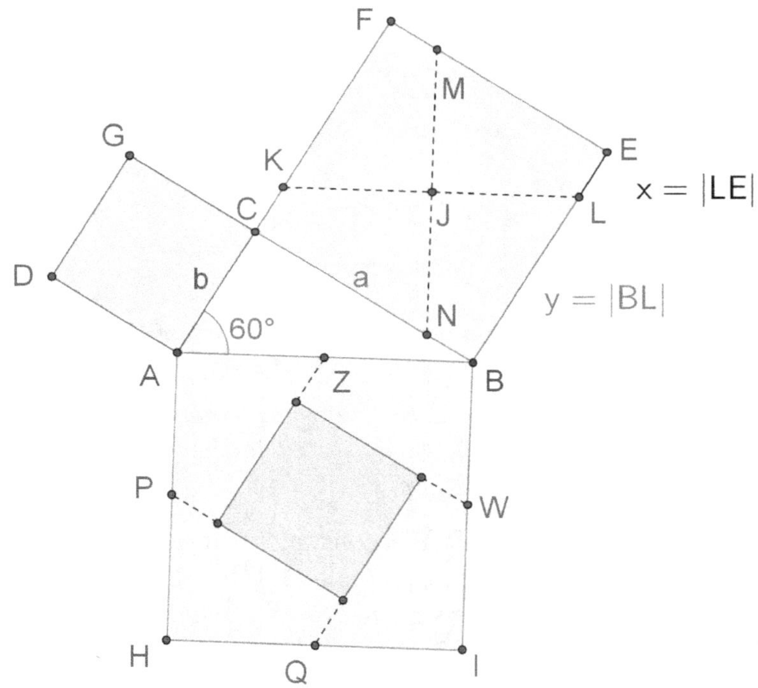

Figure IV.11.10: Dissection proof of the Pythagorean theorem in a special case

Problem IV.11.9. *Given is a right triangle $\triangle ABC$ with angle $\alpha = 60°$ and side $a = 2$. Squares are erected on its three sides. As done in the figure on page 210, one can construct dissections of them which prove the Theorem of Pythagoras. What kind of quadrilateral is $\square ALKB$. Determine the segment lengths $x = |LE|$ and $y = |BL|$.*

Answer. The quadrilateral $\square ABLK$ is a parallelogram. Its opposite sides are congruent. Hence $y = |AK| = |AC| + |CK| = |AC| + |LE| = b + x$. Secondly,

$a = |BE| = |BL| + |LE| = y + x$. From the equations

$$x + y = a\,,\ y - x = b \quad \text{one concludes}$$

$$x = \frac{a-b}{2} = 1 - \frac{\sqrt{3}}{3} \quad \text{and} \quad y = \frac{a+b}{2} = 1 + \frac{\sqrt{3}}{3}$$

IV.11.3 Dudeney's Dissection Problem

How can one dissect a unit square into a few polygons and rearrange them into an equilateral triangle? This is a problem from "The Canterbury Puzzles" by H.E. Dudeney [3], first published in 1908. Dudeney gave the solution indicated in the drawing below, which uses just four polygons. A less ingenious solution is given in

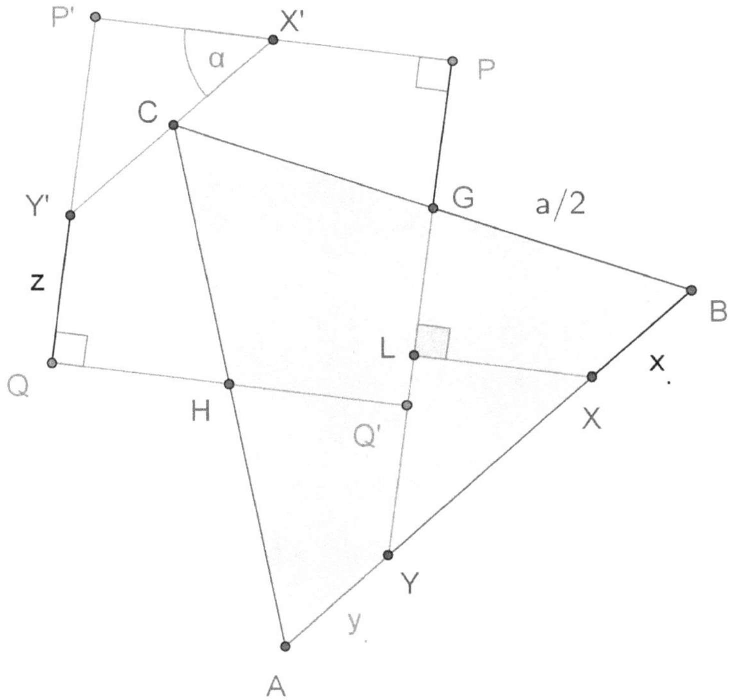

Figure IV.11.11: Dudeney's dissection of an equilateral triangle and a square of the same area.

the figure on page 213. It uses five pieces some of which occur as mirror images.

Problem IV.11.10. *We assume the side of the square to be equal to one. Calculate segments a and $x + y$ and the angle α in the accompanying drawing for Dudeney's dissection to confirm its existence.*

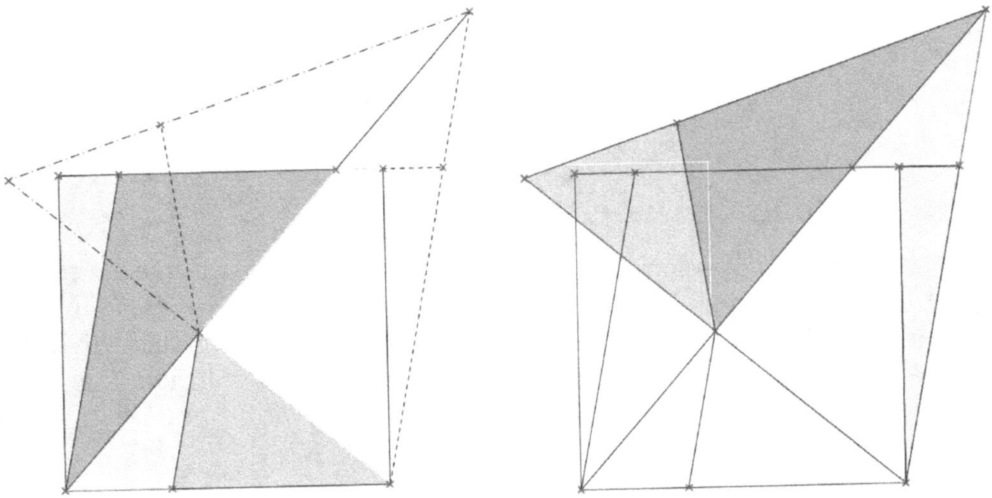

Figure IV.11.12: Another dissection of an equilateral triangle and a square of the same area.

Answer. The side a is obtained from the total area $1 = \frac{\sqrt{3}}{4} a^2$ to be

$$a = 2\,(3^{-1/4}) = 2\sqrt{\frac{\sqrt{3}}{3}}$$

The two congruent quadrilaterals on the left side

$$\square HAYQ \cong \square HCY'Q$$

are obtained rotating by 180° about H. Hence $y = |AY| = |CY'|$. The two congruent quadrilaterals on the right side

$$\square GBXL \cong \square GCX'P$$

are obtained rotating by 180° about G. Hence $x = |BX| = |CX'|$. To complete the dissection, the triangles

$$\triangle XLY \cong \triangle X'P'Y'$$

have to be congruent. Hence $a - x - y = |XY| = |X'Y'| = x + y$ and

$$x + y = \frac{a}{2}$$

Since $|P'X'| = |LX| = |X'P| = \frac{1}{2}$, we get from the right triangle $\triangle X'P'Y'$

$$\cos\alpha = \frac{|P'X'|}{|X'Y'|} = \frac{1}{a} = \frac{3^{1/4}}{2} < \frac{\sqrt{2}}{2} \quad \text{and} \quad \alpha \approx 48.85^\circ > 45^\circ$$

Problem IV.11.11. *Find four segments of the length $z = |QY'|$. Calculate the length $|Q'L|$ in terms of z.*

Problem IV.11.12. *Give a Euclidean construction for Dudeney's dissection beginning with a unit square.*

Answer. The three figures below on page 214, page 215, and page 216 give stages of such a construction.

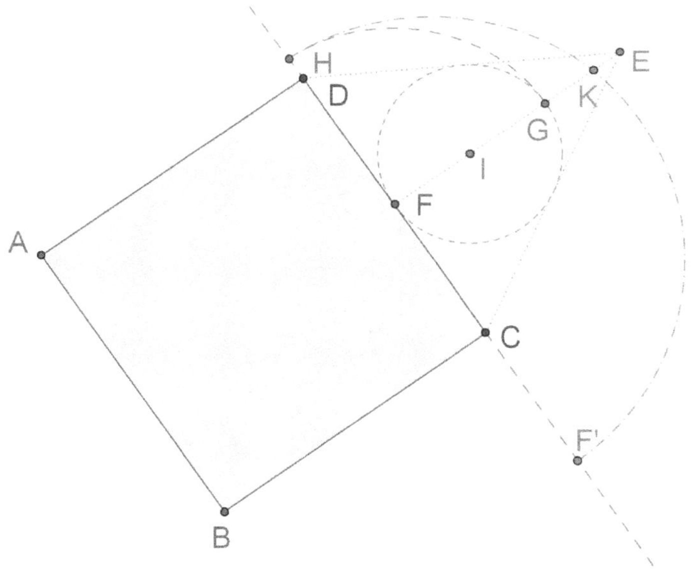

Figure IV.11.13: A construction of Dudeney's dissection starting with the square I.

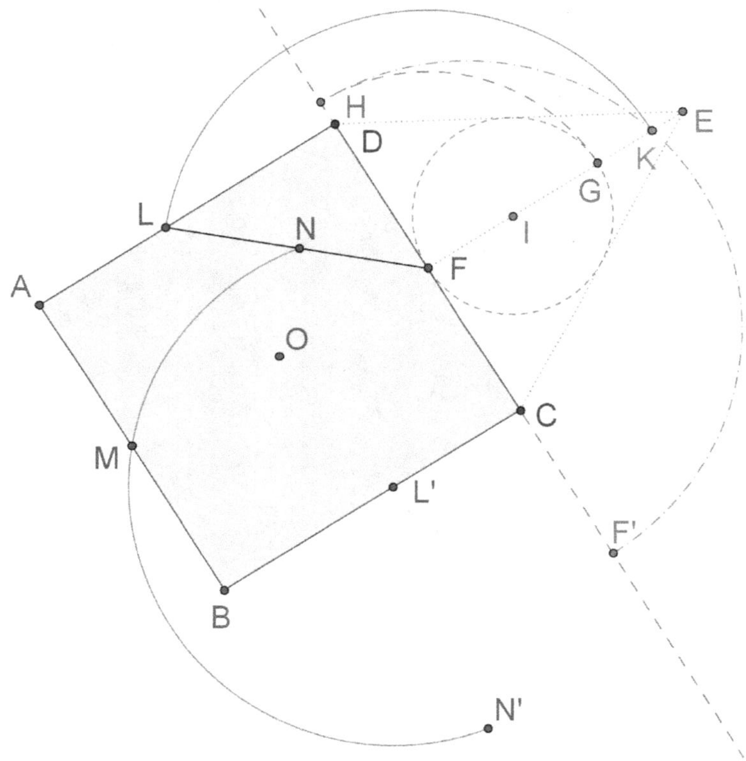

Figure IV.11.14: A construction of Dudeney's dissection starting with the square II.

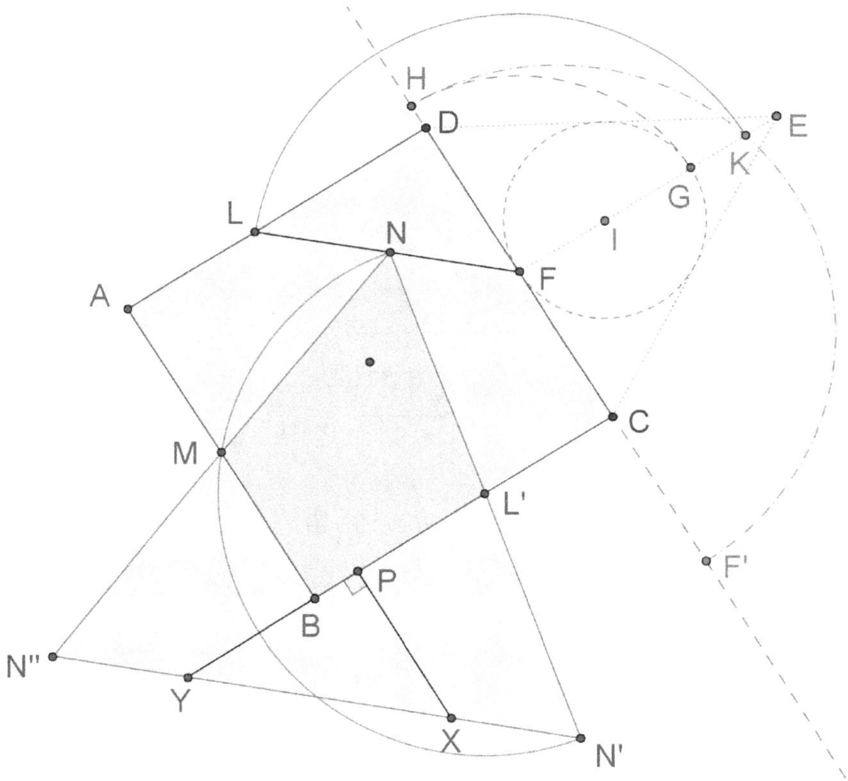

Figure IV.11.15: A construction of Dudeney's dissection starting with the square III.

Problem IV.11.13. *Calculate the segment x and the ratios $\frac{x}{a}$ and $\frac{y}{a}$. You can use the cosine theorem for the triangle $\triangle YGB$.*

Answer. The sides of the triangle $\triangle YGB$ have the lengths

$$|YG| = 1\,,\ |GB| = \frac{a}{2}\,,\ |BY| = x + \frac{a}{2}$$

Using the 60° angle at vertex B, the cosine Theorem yields

$$\begin{aligned} 1^2 = \frac{a^2}{4} + \left(x + \frac{a}{2}\right)^2 - 2\frac{a}{2}\left(x + \frac{a}{2}\right)\cos 60^\circ &= x^2 + xa + \frac{a^2}{2} - \frac{a}{2}\left(x + \frac{a}{2}\right) \\ x^2 + \frac{ax}{2} + \frac{a^2}{4} - 1 &= 0 \\ x &= \frac{-a + \sqrt{16 - 3a^2}}{4} \\ \frac{x}{a} &= \frac{-1 + \sqrt{4\sqrt{3} - 3}}{4} \\ \frac{y}{a} &= \frac{3 - \sqrt{4\sqrt{3} - 3}}{4} \end{aligned}$$

Hence

$$x \approx .24549 \quad \text{and} \quad y \approx .25451$$

Problem IV.11.14. *Give a Euclidean construction for Dudeney's dissection beginning with a given equilateral triangle.*

Answer. The figure on page 218 shows such a construction.

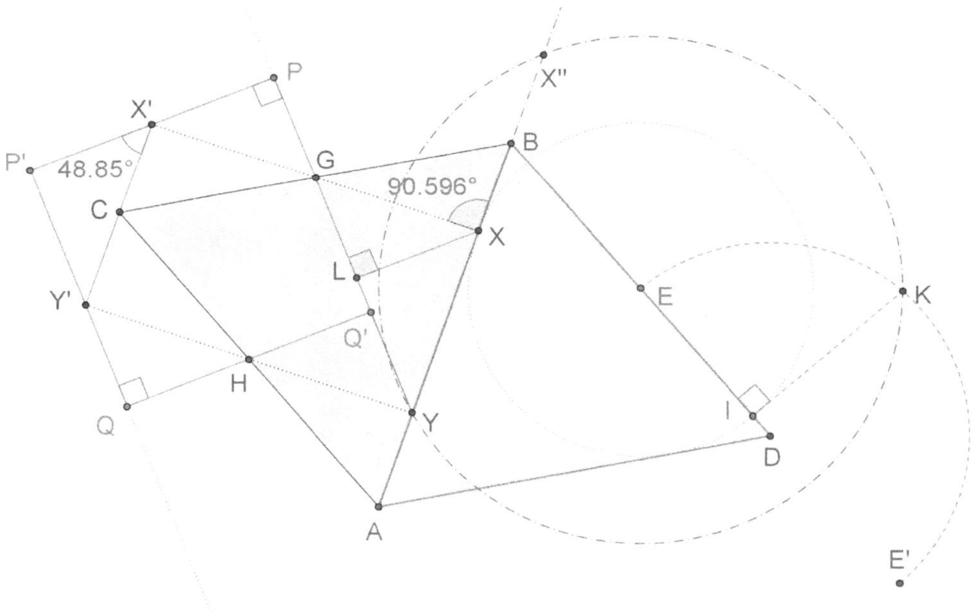

Figure IV.11.16: A construction of Dudeney's dissection starting with the triangle.

IV.11.4 Euclidean Area

As the first step, we define the area for any triangle $\triangle ABC$. In Euclidean geometry, the area of a triangle is defined to be half of base times height.

$$\text{area}(\triangle ABC) = \frac{1}{2}ah_a = \frac{1}{2}bh_b = \frac{1}{2}ch_c$$

Problem IV.11.15. *Prove that it does not matter in this definition, which side of the triangle is chosen as base.* [13]

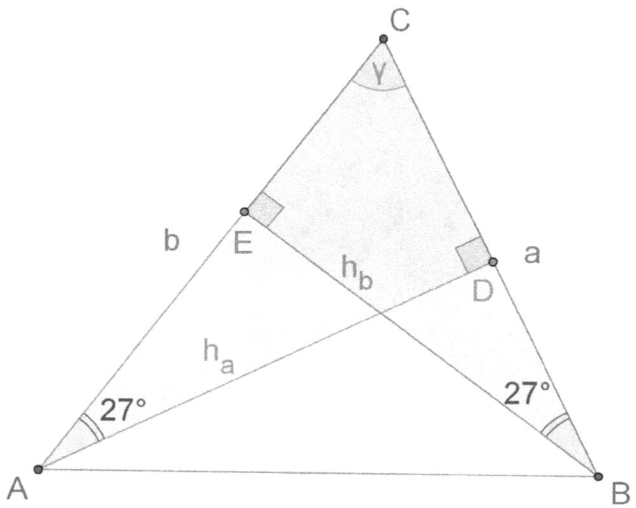

Figure IV.11.17: For the calculation of its area, any side of a triangle may be used as its base.

Answer. For the triangle $\triangle ABC$, we can take side BC as base. The corresponding altitude is AD, were D is the foot-point of the perpendicular dropped from vertex A onto side BC.

As a second possibility, we can take side AC as base. The corresponding altitude is BE, were E is the foot-point of the perpendicular dropped from vertex B onto side AC.

[13] If one wants to define the area of rectangle as width time length, one get a similar difficulty—which again needs to be resolved by means of similar triangles.

The two triangles $\triangle CAD$ and $\triangle CBE$ are equiangular, and hence similar. We get the proportion

$$\frac{h_a}{b} = \frac{|AD|}{|AC|} = \frac{|BE|}{|BC|} = \frac{h_b}{a} = \sin\gamma$$

By multiplication with the denominators, we obtain

$$ah_a = |AD| \cdot |BC| = |BE| \cdot |AC| = bh_b = \sin\gamma \cdot a \cdot b$$

which is both the double area. We see that the area of a triangle well defined to be half of base times height. It does not matter which side one chooses as base.

Remark. We see from this definition that the area takes values the field of segment arithmetic. This is indeed an Abelian group, as required. Note that this step uses the similarly of triangles with different orientation.

Proposition IV.11.9. *The area of the triangle is given by one half product of two sides time and the sin of the angle between them:*

$$\text{area}(\triangle ABC) = \frac{1}{2}ab\sin\gamma = \frac{1}{2}bc\sin\alpha = \frac{1}{2}ca\sin\beta$$

Proposition IV.11.10 (Instead of Euclid I.39)**.** *Two equicomplementable triangles with the same base have congruent altitudes.*

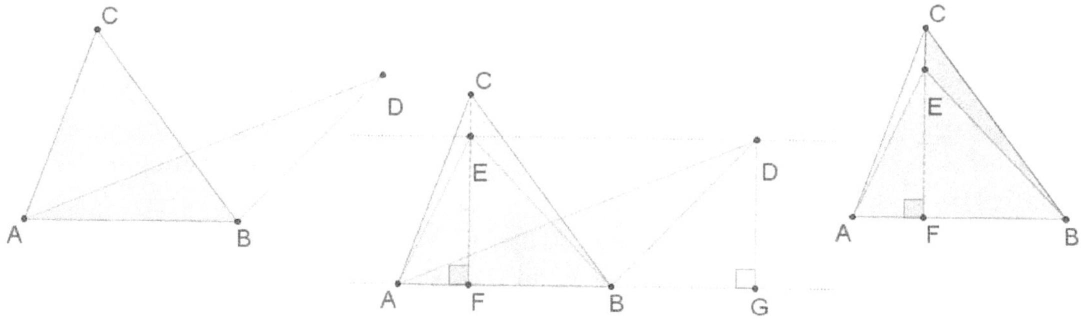

Figure IV.11.18: Two triangles with same base and different altitudes are not equicomplementable.

The proof inspired by Euclid needs to assume the following postulate.

Proposition IV.11.11 (The postulate of de Zolt and Stolz). *Let $Q \subset P$ be two figures such that there exists a triangle T inside P which does not overlap with the smaller figure Q. Then the figures P and Q are not equicomplementable.*

In the section on area in neutral geometry, this postulate is derived in Hilbert's system of axioms.

Proof inspired by Euclid. We assume that the triangles $\triangle ABC$ and $\triangle ABD$ are equicomplementable. Let their altitudes be $CF \perp AB$ and $DG \perp AB$. To check whether they are congruent, we transfer GD onto the ray $\overrightarrow{FC}$ and get the congruent segment $FE \cong GD$. The triangles $\triangle ABD$ and $\triangle ABE$ have the same base and congruent altitudes. By the proposition IV.11.5, which Euclid has already proved, they are equicomplementable. Hence by theorem IV.11.1, the triangles $\triangle ABC$ and $\triangle ABE$ are equicomplementable. Too, their altitudes dropped from C and E have the same foot point F, and lie on the same line.

The question is now, whether $C = E$. If not, there are two possible cases left. Either E lies between F and C, or C lies between F and E. It is enough to rule out the first case, the argument in the second case is similar.

Assume—towards a contradiction—that $F * E * C$. In that case the triangle $\triangle ABE$ is a subset of triangle $\triangle ABC$, to which it is equicomplementable. Too, the triangle $\triangle ECB$ lies inside $\triangle ABC$, but does not overlap the smaller triangle $\triangle ABE$. Hence, by deZolt's postulate IV.11.11, the triangles $\triangle ABE$ and $\triangle ABC$ cannot be equicomplementable. This contradiction rules out the case $F * E * C$. The case $F * C * E$ can be ruled out similarly.

Hence $E = C$, and the originally given triangles $\triangle ABC$ and $\triangle ABD$ have congruent altitudes $CF = EF \cong DG \perp AB$, as to be shown. □

Remark. Euclid does not succeed to prove deZolt's postulate in his treatment of area. But he did see that an additional assumptions is needed to get besides the simpler Euclid I.37 its converse Euclid I.39. To this end, he suggests an even stronger very general postulate about any magnitudes:

"*The whole is more than its parts*".

Such a general principle is clearly not justifiable. Indeed, it is not valid if dissection constructed from infinite sets,—obtained with the axiom of choice,—are allowed. This is exemplified by the Banach-Tarski paradox.

As we have seen from the proof above, deZolt's postulate would have been a workable version for a new axiom that Euclid could have used to complete his theory of area.

Short simple proof of Proposition IV.11.10. Let $\triangle ABC$ and $\triangle ABD$ be equicomplementable. By theorem V.12.6, they have same area. The area is given by half the product of base times height. Let $CF \perp AB$ and $DG \perp AB$ be their altitudes. Hence we know that

$$\frac{1}{2}|AB| \cdot |CF| = \frac{1}{2}|AB| \cdot |DG|$$

Since segment arithmetic is a field, we can multiply both sides by 2 and divide by $|AB| \neq 0$ and get

$$|CF| = |DG|$$

from which follows $CF \cong DG$, as to be shown. □

Already in the segment arithmetic, it was necessary to introduce a *unit segment.* Areas are naturally measured by taking the square over the unit segment as unit for area.

Lemma IV.11.1. *A right triangle T with legs of lengths a and b is equicomplementable to a right triangle with legs 1 and ab, and to a rectangle with legs 1 and* area(T).

Problem IV.11.16. *Prove lemma IV.11.1.*

Proposition IV.11.12. *Every figure P is equicomplementable to a rectangle one side of which is the unit segment. The second side of this rectangle has the length* area(P).

Proof. The given figure P can be dissected into triangles. Each of them is either a right triangle or can be dissected into two right triangles. Hence the figure P can be dissected into finitely many right triangles T_i for $i = 1...n$. Each of them is equicomplementable to a rectangle R_i with sides 1 and area(T_i). By proposition IV.11.1,

$$P = \biguplus_{i=1}^{n} T_i \triangleq \biguplus_{i=1}^{n} R_i = R$$

were the latter is a rectangle R with sides 1 and

$$\sum_{i=1}^{n} \text{area}(T_i) = \text{area}(P)$$

as to be shown. □

Theorem IV.11.4. *Any two figures with same area are equicomplementable.*

Proof. Let P and Q be two figures with same area area(P) = area(Q). By proposition IV.11.12, figure $P \triangleq R$, where R is a rectangle, one side of which is the unit segment, and the other side has the length area(P). Similarly, figure $Q \triangleq R'$, where

R' is a rectangle with lengths of sides 1 and area(Q). But two rectangles with the same lengths of sides are congruent, and hence clearly $R \triangleq R'$. By theorem IV.11.1, "equicomplementable" is an equivalence relation. Hence

$$P \triangleq R \triangleq R' \triangleq Q$$

implies the originally given figures P and Q are equicomplementable, as to be shown. □

Corollary 21. *Any two figures are equicomplementable if and only if they have equal areas.*

IV.11.5 The Role of the Archimedean Axiom

Theorem IV.11.5. *Assuming the Archimedean axiom* (V.1) *holds, any two figures with same area are even equidecomposable.*

Problem IV.11.17. *We assume that the Archimedean axiom holds. Explain why a right triangle T is legs of lengths a and b is even equidecomposable to a rectangle with legs* 1 *and* area(T).

Proof of Theorem IV.11.6. As explained in the problem IV.11.17 above, with the assumption of the Archimedean axiom, a right triangle T with legs of lengths a and b and a rectangle with legs 1 and area(T) are shown even to be equidecomposable.

In the same way as in proposition IV.11.12, we show that any figure P is even equidecomposable to a rectangle one side of which is the unit segment. The second side of this rectangle has the length area(P).

Finally, as in the proof of theorem IV.11.4 that any two figures P and Q with same area are equidecomposable. (We do not use the Archimedean axiom in the two last steps, but the reasoning is done now with the stronger equivalence relation "equidecomposable".) □

Problem IV.11.18. *Between two parallels $AB \parallel CD$ of distance* 5 *units, there are given two parallelograms $\square ABCD$ and $\square ABEF$. Let the lengths of side be* $|AB| = 1$, $|AC| = 10$ *and* $|AE| = 20$.

Construct dissections into triangles to shown that the two parallelograms are equidecomposable. How many triangles do you need?

Problem IV.11.19. *Let n be a natural number. What is the number t of triangles which are needed to get congruent dissections of a square of side n and a rectangle of sides* 1 *and* n^2?

(a) *Prove that at least* $\ulcorner\sqrt{2}\,n\urcorner$ *triangles are needed.*

(b) *Give a dissection using* $2n$ *triangles.*

Can one improve the lower or upper bound?

Answer. Suppose we have congruent dissections of the square of side n and rectangle of sides 1 and n^2 into t triangles. Let L be the length of the longest side of these triangles.

The two opposite long sides of the rectangle are dissected into sides of different triangles. Since each triangle fits into the square of side n, the longest side L can have length at most $\sqrt{2}\,n$. Hence

$$2n^2 \leq t \cdot L \leq t\sqrt{2}\,n$$

which implies $t \geq \ulcorner\sqrt{2}\,n\urcorner$ as claimed.

Proposition IV.11.13. *Assume the Archimedean axiom* (V.1) *does* <u>*not*</u> *hold. Then there exist a square and a rectangle that are equicomplementable but* <u>*not*</u> *equidecomposable.*

Proof. We assume the Archimedean axiom (V.1) does not hold. Hence there exist segments OB and OA such that

$$OA > n \cdot OB \quad \text{for all natural numbers } n$$

Now OB takes the role of the unit segment. We put the right triangle $\triangle OAB$ and the right isosceles triangle $\triangle OAA'$ on the same side of their common leg OA. We draw the parallel to the hypothenuse AB through point A' and get an intersection point P with the ray $\overrightarrow{OA}$. In this way, we have constructed a segment OP such that

$$|OP| \cdot |OB| = |OA|^2$$

The right triangle $\triangle OBP$ and the right isosceles triangle $\triangle OAA'$ are equicomplementable. By doubling these figures, we get a rectangle with sides OB and OP, as well as a square with side OA, which, too, are equicomplementable.

We claim these two figures are <u>not</u> equidecomposable. Suppose towards a contradiction we have congruent dissections of the square and the rectangle into t triangles. Let L be the length of the longest side of these triangles.

The two opposite long sides of the rectangle together have total length $2|OP|$. They are dissected into sides of different triangles. Since each triangle fits into the square, the longest side L can have length at most $\sqrt{2}\,|OA|$. Hence

$$\begin{aligned} 2|OP| &\leq t \cdot L \leq t\sqrt{2}\,|OA| \\ \sqrt{2}|OP| \cdot |OA| &\leq t|OA|^2 = t|OP| \cdot |OB| \\ \sqrt{2}|OA| &\leq t|OB| \end{aligned}$$

On the other hand, we have assumed $OA > n \cdot OB$ for all natural numbers n, and hence with $n := t$ we get

$$t|OB| < |OA|$$

Thus we arrive at the contradiction $\sqrt{2}|OA| < |OA|$. This confirms the claim the square and the rectangle are <u>not</u> equidecomposable. □

The logical relations between the different concepts of content gained in this section are summed up in the following diagram:

$$\begin{array}{ccc}
P \text{ and } Q \text{ are equidecomposable} & \xrightarrow{\text{clear}} & P \text{ and } Q \text{ are equicomplementable} \\
\big\downarrow \text{Th. V.12.6} & & \text{Th. V.12.6} \big\downarrow \\
\text{area}(P) & = & \text{area}(Q) \\
\big\downarrow \text{Archimedes and Th. IV.11.6} & & \text{Th. IV.11.4} \big\downarrow \\
P \text{ and } Q \text{ are equidecomposable} & \xleftarrow[\text{Archimedean axiom}]{} & P \text{ and } Q \text{ are equicomplementable}
\end{array}$$

Corollary 22. *Assuming the Archimedean axiom* (V.1) *holds, any two equicomplementable figures are even equidecomposable. Assuming the Archimedean axiom* (V.1) *does <u>not</u> hold, there exist two equicomplementable figures that are <u>not</u> equidecomposable.*

Problem IV.11.20. *Convince yourself once more that two parallelograms with the same base and congruent altitudes are always equicomplementable. Why can one not claim they are always equidecomposable?*

IV.11.6 Some Uniqueness Results for Justification

Given is a Pythagorean plane. Let UQ denote a square the sides of unit length. Does assigning the area 1 to this figure determine the area function uniquely?

Proposition IV.11.14. *Take any Pythagorean plane. Let G be any ordered Abelian group. Suppose that μ is an measure function having the following properties:*

(1) *the measure $\mu(P) \geq 0$ of any figure P is a nonnegative value in the ordered Abelian group G;*

(2) $\mu(T) = \mu(S)$ *for any two congruent triangles $T \cong S$;*

(3) $\mu(P \uplus Q) = \mu(P) + \mu(Q)$ *for any non-overlapping figures P and Q.*

Then there exists an order preserving homomorphism $\phi : F \mapsto G$ from field F of segment arithmetic to the group G such that $\mu = \phi \circ \text{area}$. In other word, each measure is a homomorphic image of the area:

$$\mu(P) = \phi(\text{area}(P)) \quad \textit{for all figures } P \tag{IV.11.1}$$

and we get the diagram

$$\begin{array}{ccc} & \textit{figure } P & \\ \text{area} \downarrow & & \downarrow \textit{measure } \mu \\ \text{area}(P) & \xrightarrow[\phi]{} & \textit{measure } \mu(P) \end{array}$$

Proposition IV.11.15. *Under the additional assumption that*

(1*) *the measure $\mu(T) > 0$ is strictly positive for any triangle T.*

the mapping ϕ is injective, hence an endomorphism.

Proof of Proposition IV.11.14. By the same reasoning as used in the proof of Theorem V.12.6, we derive from the assumptions (2) and (3) the following more general properties of the measure μ:

(4) Any two equidecomposable figures $P \overset{\boxtimes}{\sim} Q$ have the same measure $\mu(P) = \mu(Q)$.

(5) Any two equicomplementable figures $P \triangleq Q$ have the same measure $\mu(P) = \mu(Q)$.

We go on to construct the homomorphism $\phi : F \mapsto G$ with the property (IV.11.1). For any segment length r, we define $\phi(r) = \mu(R)$ to be the measure of the rectangle R, one side of which is the unit segment and the second side has the length r.

Given is any figure P. By Proposition IV.11.12, the figure P is equicomplementable to a rectangle $R \triangleq P$, one side of which is the unit segment and the second side has the length area(P). Hence we get both area(R) = area(P), as well as $\mu(R) = \mu(P)$ from property (5). In the end that means

$$\mu(P) = \mu(R) = \phi(r) = \phi(\text{area}(R)) = \phi(\text{area}(P))$$

holds for any figure P, as required.

It remains to check that the mapping ϕ preserves the addition and the order. Take two rectangles R with sides $r \times 1$, and S with sides $s \times 1$. We may assume they are non overlapping and have a common side of unit length. In that case $R \uplus S$ is a rectangle with sides $r + s$ and 1, and thus

$$\phi(r + s) = \mu(P \uplus R) = \mu(R) + \mu(S) = \phi(r) + \phi(s)$$

confirms the additivity. Now assume that $r \leq s$, and both R and S lie on the same side of their common side of unit length. In this case $R \subseteq S$, and hence

$$\phi(r) = \mu(R) \leq \mu(S) = \phi(s)$$

confirms that ϕ preserves the order. □

Proof of Proposition IV.11.15. In order to check whether the homomorphism ϕ is injective, we suppose that r and s are two segment lengths and $\phi(r) = \phi(s)$. The rectangles R with sides $r \times 1$ and S with sides $s \times 1$ have the same measure since $\mu(R) = \phi(r) = \phi(s) = \mu(S)$. Since any two segments are comparable, we may assume $r \leq s$ and hence $R \subseteq S$.

We use the equality $\mu(R) = \mu(S)$ to rule out the R is strictly less that S. In that case, the difference $S \setminus R$ would be the disjoint union of two congruent right triangles. By assumption (1*), they would have a positive measure and hence additivity would imply $\mu(R) < \mu(S)$.

Thus we have seen that $\phi(r) = \phi(s)$ implies $\mu(R) = \mu(S)$, secondly $R = S$ and finally $r = s$, as to be checked for injectivity of ϕ. □

Theorem IV.11.6. *In a Pythagorean and* <u>*Archimedean*</u> *plane, any measure function μ is equal to the common known* area, *provided it has the following properties:*

(0) *the measure $\mu(UQ) = 1$ for a unit square;*

(1) *the measure $\mu(P) \geq 0$ of any figure P is a nonnegative value in the segment arithmetic;*

(2) *$\mu(T) = \mu(S)$ for any two congruent triangles $T \cong S$;*

(3) $\mu(P \uplus Q) = \mu(P) + \mu(Q)$ *for any non-overlapping figures* P *and* Q.

It is interesting to note that this uniqueness proof uses really Eudoxus idea to get equality from rational upper and lower bounds. This principle appear in Euclid's book V as a definition for the equality of ratios of magnitudes. We reformulate this result as a proposition about Archimedean fields.

Proposition IV.11.16 (Eudoxus principle). *In an Archimedean field are given any four elements* $a, b, c, d > 0$. *The ratios* $a : b$ *and* $c : d$ *are equal, if and only if for any whole numbers* m *and* n *either one of these three cases occurs:*

both $na > mb$ *and* $nc > md$,

both $na = mb$ *and* $nc = md$,

both $na < mb$ *and* $nc < md$.

Proof. By the the Archimedean axiom there exists a natural number n such that $(n+1) \cdot a > m \cdot b$. By the basic properties of the natural numbers, there exists a *smallest* such n. In this case both $n \cdot a \leq m \cdot b$ as well as $(n+1) \cdot a > m \cdot b$ hold. By assumption, the same inequalities hold for a and b replaced by c and d. Hence

$$n \cdot a \leq m \cdot b < (n+1) \cdot a \quad \text{and} \quad n \cdot c \leq m \cdot d < (n+1) \cdot c$$

Division implies

$$\frac{n}{m} \leq \frac{b}{a} < \frac{n+1}{m} \quad \text{and} \quad \frac{n}{m} \leq \frac{c}{d} < \frac{n+1}{m}$$

We have squeezed both fractions b/a and c/d
<u>between the same upper and lower rational bounds</u>. By subtraction we conclude

$$-\frac{1}{m} \leq \frac{b}{a} - \frac{c}{d} \leq \frac{1}{m}$$

Since m is an arbitrary natural number, and the Archimedean axiom has been assumed to hold, we conclude $\frac{b}{a} = \frac{c}{d}$. This last step also called the *principle of exhaustion.* □

Remark. Eudoxus and Archimedes' principles are used by Euclid in book XII.2 to prove that the *"The areas of circles are in the same ratio as the squares on their diameters"*. Too, they are essential for Euclid's theory of volume. Lateron, the same principle appear again in Archimedean proof of the law of the lever, and many times more in mathematical physics.

Proof of Proposition IV.11.6 . By Proposition IV.11.14, there exists a homomorphism $\phi : F \mapsto F$ from the field of segment arithmetic to itself such that

$$\mu(P) = \phi(\text{area}(P)) \quad \text{for all figures } P \tag{IV.11.1}$$

Moreover normalization for the unit square implies

$$\phi(1) = \phi(\text{area}(UQ)) = \mu(1) = 1$$

The additivity of the area and the measure of any n non overlapping congruent rectangles R of dimensions $r \times 1$ imply, by means of an easy induction on n that

$$\begin{aligned}\phi(n \cdot r) &= \phi(\text{area}(n \cdot R)) = \mu(n \cdot R) \\ &= n \cdot \mu(R) = n \cdot \phi(\text{area}(R)) = n \cdot \phi(r)\end{aligned}$$

For any natural n and m we conclude

$$\begin{aligned}m \cdot \phi\left(\frac{r}{m}\right) &= \phi\left(m \cdot \frac{r}{m}\right) = \phi(r) \\ \phi\left(\frac{1}{m} \cdot r\right) &= \frac{1}{m} \cdot \phi(r) \\ \phi\left(\frac{n}{m} \cdot r\right) &= \frac{n}{m} \cdot \phi(r)\end{aligned}$$

proving $\phi(qr) = q\,\phi(r)$ for any rational q. We need the corresponding property for any even irrational segment length s.

Given is any segment $s > 0$ and denominator $m \neq 0$, by the Archimedean axiom there exists a natural number n such that $(n+1) \cdot 1 > m \cdot s$. By the basic properties of the natural numbers, there exists a *smallest* such n. In this case, we get

$$n \cdot 1 \leq m \cdot s < (n+1) \cdot 1 \quad \text{and} \quad n \cdot 1 \leq m \cdot \phi(s) < (n+1) \cdot 1$$

The second inequalities hold since $\phi(qr) = q\phi(r)$ for any rational q and $r = 1$, and $\phi(1) = 1$, and finally the homomorphism ϕ preserves the order. We have squeezed both s and $\phi(s)$ between the same upper and lower rational bounds, and conclude by subtraction

$$-\frac{1}{m} \cdot 1 \leq \phi(s) - s \leq \frac{1}{m} \cdot 1$$

Since m is an arbitrary natural number, and the Archimedean axiom has been assumed to hold, we conclude $\phi(s) = s$. Once more, we have used Archimedes' *principle of exhaustion.* Hence it turns out that ϕ is the identity and the only additive positive measure μ is the common area. □

IV.11.7 About the Volume of Polyhedra

Already Gauss brought it to the attention of mathematicians, whether a corresponding justification of the three-dimensional volume based on dissections is possible. In a letter to Gerling in 1844, Gauss says that it is too bad that the equality of volume of two symmetrical, but not congruent figures, can be proved only using the method of exhaustion. Gerling could solve the dissection problem for this example. In his reply, Gerling gives a direct proof that any triangular pyramid can be dissected into twelve pieces that are congruent to those of a dissection of its mirror image. Gauss responds that this is a nice result. But it is still unfortunate that the proof of Euclid's proposition (XII.5)

`"Pyramids of the same height on triangular bases are in the same ratio as their bases"`

needs the method of exhaustion. Hilbert conjectured that the method of exhaustion is really necessary. At the International Congress of Mathematicians in Paris 1900, he presented his famous list of the 23 most important problems facing mathematics in the twentieth century. The third problem of this list addresses the present question:

> Give two tetrahedra with equal bases and equal heights, which in no way can be dissected into congruent tetrahedra, and which, even by adding congruent tetrahedra, cannot be complemented into polyhedra, which in turn can be dissected into congruent tetrahedra.

In the same year, Max Dehn has given such a counterexample. Max Dehn proves that it is not possible to dissect a regular tetrahedron into a finite number of pieces that can be reassembled into a cube, even after possibly adding on other figures that are equivalent by dissection. The important consequence is that the method of exhaustion is really necessary in Euclid's proof of (XII.5).

IV.11.8 Heron's Formula for the Area of a Triangle

Problem IV.11.21. *Derive Heron's formula*

$$\text{area}(\triangle ABC) = \frac{1}{4}\sqrt{(a+b+c)(a+b-c)(a-b+c)(b+c-a)}$$

for the area a triangle in terms of its three sides a, b and c.

Answer. Take the triangle $\triangle ABC$ with vertices at the centers of the two circles and their intersection point A. We begin with the basic fact that the area of a triangle is

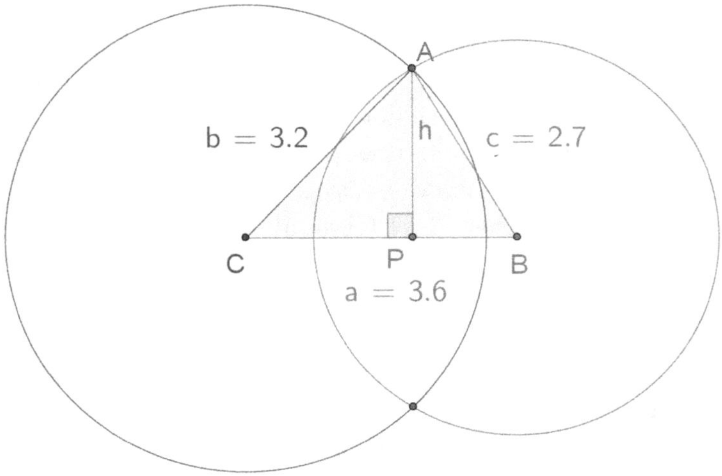

Figure IV.11.19: Herons' formula for the area of triangle.

half the product of base and altitude. We use the base CB and the altitude AP and get

$$\text{Area } \triangle ABC = \frac{1}{2}\,|CB| \cdot |AP|$$

Let $|CP| = p$ and $|BP| = q$. The height is calculated with the Pythagorean theorem in the right triangle $\triangle CPA$

$$h^2 = |AP|^2 = |CA|^2 - |CP|^2 = b^2 - p^2 = (b-p)(b+p)$$

Similarly, the Pythagorean theorem in the right triangle $\triangle BPA$ yields

$$h^2 = |AP|^2 = |BA|^2 - |BP|^2 = c^2 - q^2$$

Subtracting yields

$$\begin{aligned} b^2 - c^2 &= p^2 - q^2 = p^2 - (a-p)^2 \\ a^2 + b^2 - c^2 &= a^2 + p^2 - (a-p)^2 = 2ap \\ 2a(b+p) &= (a+b)^2 - c^2 = (a+b+c)(a+b-c) \\ 2a(b-p) &= -(a-b)^2 + c^2 = (-a+b+c)(a-b+c) \end{aligned}$$

Hence we get

$$\begin{aligned}\text{Area } \triangle ABC &= \frac{1}{2}\,|CB|\cdot|AP| = \frac{1}{4}\,\sqrt{2a(b+p)}\,\sqrt{2a(b-p)}\\ &= \frac{1}{4}\,\sqrt{(a+b+c)(a+b-c)(-a+b+c)(a-b+c)}\end{aligned}$$

IV.11.9 Algebraic Relations for the Pieces of a Triangle

Problem IV.11.22. *Let r be the radius of the in-circle of triangle $\triangle ABC$. Prove the area is*

$$\text{area}(\triangle ABC) = \frac{(a+b+c)\,r}{2}$$

Answer. It is explained in an earlier problem V.12.4 in the context of neutral triangle geometry, how to construct the in-circle of a triangle. The perpendiculars are dropped

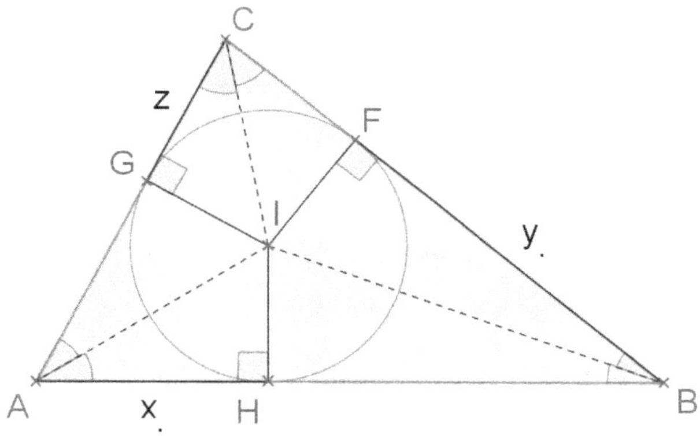

Figure IV.11.20: An arbitrary triangle is partitioned into six right triangles.

from its center I onto the three sides. The foot points F, G, H are the touching points of the in-circle to the three sides. The six segments from the center of the in-circle to the vertices and the touching points partition the given triangle into six right triangles. All of them have the radius of the in-circle as one of their legs. The other legs have lengths $x = |AH|$, $y = |BF|$, and $z = |CG|$, each one occurring twice for two congruent triangles. Since the circumference is $a+b+c = 2x+2y+2z$, the total area of the given triangle $\triangle ABC$ is

$$\text{area}(\triangle ABC) = 2\cdot\frac{x\,r}{2} + 2\cdot\frac{y\,r}{2} + 2\cdot\frac{z\,r}{2} = (x+y+z)\,r = \frac{(a+b+c)\,r}{2}$$

Problem IV.11.23. *Let R is the radius of the circum-circle of triangle $\triangle ABC$. Prove the area is*

$$\operatorname{area}(\triangle ABC) = \frac{abc}{4R}$$

Answer. As already stated in the Proposition IV.11.9, the area of a triangle is one half product of two sides time and the sin of the angle between them:

$$\operatorname{area}(\triangle ABC) = \frac{1}{2} ab \sin \gamma$$

On the other hand, the extended sin theorem tells that

$$\frac{c}{\sin \gamma} = 2R$$

Hence $\sin \gamma = \frac{c}{2R}$. Plugging into the first formula yields the required formula

$$\operatorname{area}(\triangle ABC) = \frac{abc}{4R}$$

Problem IV.11.24. *Can the three sides of a triangle be determined from the radius of the circum-circle, the radius of the in-circle and the circumference of the triangle?*

Answer. Yes, this is possible, but one needs to solve a third order equation. Hence the problem cannot be solved by a Euclidean construction— one needs to use Cardanos' formula.

We assume as given: the radius r of the in-circle, the radius R of the circum-circle, and the circumference $2s$. From the two previous problems, we get

$$\operatorname{area}(\triangle ABC) = \frac{abc}{4R} = s\,r$$

Hence we can calculate the sum $a + b + c = 2s$ and the product $abc = 4Rrs$. We define the polynomial

$$P(x) := (x-a)(x-b)(x-c) = x^3 - (a+b+c)x^2 + (ab+ac+bc)x - abc$$

To get all coefficients of this polynomial, we have still to use Heron's formula, given in Problem IV.11.21 below in the form

$$\operatorname{area}(\triangle ABC) = \frac{1}{4}\sqrt{(a+b+c)(a+b-c)(a-b+c)(b+c-a)}$$

This can also be written with the semi-perimeter s as

$$\text{area}(\triangle ABC) = \sqrt{s(s-a)(s-b)(s-c)}$$

Hence we get one further value of the polynomial P:

$$P(s) = \frac{\text{area}(\triangle ABC)^2}{s} = rs^2$$

One now can calculate the polynomial and obtain

$$P(x) = x^3 - 2sx^2 + (rs + s^2 + 4Rr)x - 4Rrs$$

From this information, it is possible to calculate the three sides, for example by Cardanos' formula or by some numerical method.

Part V

Some Background

V.12 Citations from the First Volume

Proposition V.12.1 (The Crossbar Theorem). *A segment with endpoints on the two sides of an angle and a ray emanating from its vertex into the interior of the angle intersect.*

Theorem V.12.1 (Orientation Theorem). *In any given plane, one can assign a left and a right half plane to every oriented line in a consistent way such that*

- *reversing the orientation of a line switches its left and right half plane;*
- *any point in the interior of an angle lies in the left half plane of one of its sides, and the right half plane of the other side of the angle.*

Hence there exist exactly two possibilities to assign a left and a right half plane to all oriented lines in a plane.

Proposition V.12.2 (Restricted Jordan Curve Theorem). *A simple closed polygonal curve $\mathcal{C}$ partitions the points of the plane not lying on the polygon into two regions, called the* interior *and* exterior *of the polygon, which have the following properties:*

- *If A is a point of the interior, and B a point of the exterior region, then each polygonal curve con9o090onecting A to B has at least one point in common with the curve $\mathcal{C}$.*
- *If A and A' are two points of the interior region, then there exists a polygonal curve connecting A to A' which has no point in common with the curve $\mathcal{C}$.*
- *Similarly, if B and B' are two points of the exterior region, there exists a polygonal curve connecting B to B' which has no point in common with the curve $\mathcal{C}$.*
- *There exists a line which does not cut the interior region.*
- *Every line cuts the exterior region.*

Proposition V.12.3 (Existence and uniqueness of segment transfer). *Given a segment AB and given a ray r originating at point A', there exists a unique point B' on the ray r such that $AB \cong A'B'$.*

Definition V.12.1 (Sum of segments). The sum of any two segments AB and EF is defined to be the segment AC were C is the point such that $A * B * C$ and $EF \cong BC$.

Proposition V.12.4 (Congruence classes of segments are an ordered Abelian group). *The sum of segments is defined on equivalence classes of congruent segments because of axiom* (III.3) *and the existence and uniqueness of segment transfer, as shown in Proposition V.12.3. These equivalence classes have the following properties:*

Commutativity $a + b = b + a$.

Associativity $(a + b) + c = a + (b + c)$

Comparison *Any two segments a and c satisfy either $a < c$, or $a = c$ or $a > c$.*

Difference *$a < c$ if and only if there exists b such that $a + b = c$.*

Comparison of Sums *If $a < b$ and $c < d$, then $a + c < b + d$.*

Put together with their negatives $-a$ for each congruence class a, and finally the neutral element 0, *these equivalence classes are an* ordered Abelian group.

Definition V.12.2 (Hilbert tools). By incidence axiom (I.1) and (I.2), we can draw a unique line between any two given points. By axiom (III.1) and the uniqueness proved earlier, we transfer a given segment uniquely to a given ray. Finally by axiom III.4, we transfer a given angle uniquely on a given ray into the specified half plane. Constructions done using <u>only</u> these means are called constructions by *Hilbert tools.*

Proposition V.12.5 (Congruence of Supplementary Angles). [Theorem 14 of Hilbert] *If an angle $\angle ABC$ is congruent to another angle $\angle A'B'C'$, then its supplementary angle $\angle CBD$ is congruent to the supplementary angle $\angle C'B'D'$ of the second angle.*

Proposition V.12.6 (Comparison of supplements). *If $\alpha < \beta$, then their supplements $S(\alpha)$ and $S(\beta)$ satisfy $S(\alpha) > S(\beta)$.*

Definition V.12.3 (Alternate interior angles or z-angles). Let a transversal t intersect two lines a and b at the points A and B. A pair of *alternate interior angles* or simply *z-angles* are two angles

$$\alpha := \angle(\overrightarrow{AB}, r_A) \quad \text{and} \quad \beta := \angle(\overrightarrow{BA}, r_B)$$

They have vertices A and B. One of their respective sides are the rays $r_A \subseteq a$ and $r_B \subseteq b$ lying on the lines a respectively b, and lying in opposite half-planes of the transversal t. Their other sides are the rays $\overrightarrow{AB}$ and $\overrightarrow{BA}$, which both contain the segment AB.

Definition V.12.4 (Exterior angle of a triangle). The *exterior angle of a triangle* is the supplement of an interior angle.

Proposition V.12.7 (The Exterior Angle Theorem). [Euclid I.16. Theorem 22 in Hilbert] *The exterior angle of a triangle is greater than both nonadjacent interior angles.*

Proposition V.12.8. [Euclid I.17] *The sum of any two interior angles of a triangle is less than two right angles.*

Proposition V.12.9 (Comparison of sides implies comparison of angles). [Euclid I.18, Theorem 23 of Hilbert] *In any triangle, across the longer side lies the greater angle.*

Proposition V.12.10 (Comparison of angles implies comparison of sides). [Euclid I.19] *In any triangle, across the greater angle lies the longer side.*

Proposition V.12.11 (Extended ASA-Congruence). *Given is a triangle, a segment congruent to one of its sides, and a half-plane $\mathcal{H}'$ bounded by this segment. The two angles at the vertices of this side are transferred to the endpoints of the segment, and reproduced in the same half plane. Then the newly constructed rays intersect, and one gets a second congruent triangle.*

Proposition V.12.12 (Addendum to the extended ASA-Congruence). *Given is a triangle $\triangle ABC$, a* shorter *segment $A'B' < AB$ and a half-plane $\mathcal{H}'$ bounded by this segment. The two angles at the vertices A and B are transferred to the endpoints of the segment $A'B'$, and reproduced in the same half plane $\mathcal{H}'$. Then the newly obtained rays r'_A and r'_B intersect, say at point C'. One obtains a triangle $\triangle A'B'C'$, lying in the half planen $\mathcal{H}'$, and having at vertices A' and B' the angles $\angle B'A'C' \cong \angle BAC$ and $\angle A'B'C' \cong \angle ABC$.*

Lemma V.12.1 (Lemma for SSS Congruence). [Euclid I.7] *Assume that the two triangles $\triangle ABC$ and $\triangle AB'C$ have a common side AC, and all three corresponding sides are congruent, and the two vertices B and B' lie on the same side of line AC. Then the two triangles are identical.*

Proposition V.12.13 (SSS Congruence). [Theorem 18 in Hilbert's Foundations] *Two triangles with three pairs of congruent sides are congruent.*

Corollary 23 (Restricted SSA Congruence Theorem). *Two triangles with two pairs of congruent sides, and a pair of congruent angles across one of these congruent sides are congruent if additionally*

- *both triangles are acute or right;*
- *both triangles are obtuse* and *their obtuse angles are correspondig pieces.*

Proposition V.12.14 (Existence of a parallel). *For every line l and for every point P lying not on l, there exists at least one parallel m to l through point P.*

Proposition V.12.15 (Congruent z-angles imply parallels). [Euclid I.27] *If two lines form congruent z-angles with a transversal, they are parallel.*

Proposition V.12.16 (Characterization of the perpendicular bisector). *A point P lies on the perpendicular bisector p of a segment AB if and only if it has congruent distances to both endpoints of the segment.*
Point P lies on the same side of the perpendicular bisector p of a segment AB as point B if and only if $|PB| < |PA|$.

Theorem V.12.2 (The Archimedean property for angles, Hilbert's Proposition 34). *Assume that the Archimedean Axiom holds. For any two angles α and ε there exists a natural number r such that*

$$\frac{\alpha}{2^r} < \varepsilon \tag{V.12.1}$$

Main Theorem 5 (Measurement of segments). *Given is a* unit segment *OI. There exists a unique way of assigning a* length *to any segment. We denote the length of segment AB by $|AB|$.* [14] *We construct the length as an (finite or infinite) binary fraction. Too, this length is called the* distance *of points A and B. The length has the following properties:*

(1) Positivity *If $A \neq B$, then $|AB|$ is a positive real number, and $|OI| = 1$.*

(2) Congruence *$|AB| = |CD|$ if and only if $AB \cong CD$.*

[14] In projective geometry and other contexts, one uses a signed length, which I shall denote by $\overline{AB}$.

(3) Ordering *$|AB| < |CD|$ if and only if $AB < CD$.*

(4) Additivity *The distances are additive for any three points on a line. With point B lying between A and C, we get $|AC| = |AB| + |BC|$.*

Definition V.12.5 (Short and long arc). Given is a circle with center O and a chord AB that is not a diameter. The *long arc* is the part of the circle lying in the *same* half plane of line AB as the center of the circle. The *short arc* is the part of the circle lying in the *opposite* half plane of line AB as the center of the circle.

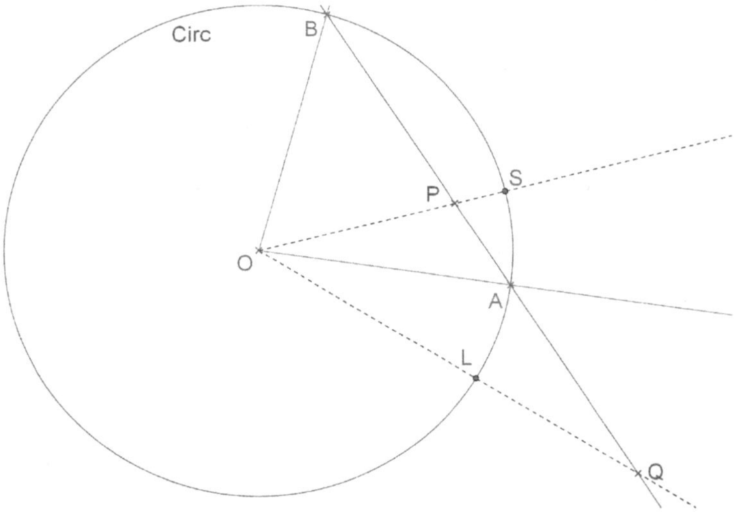

Figure V.12.1: The short/long arc lies in the interior/exterior of the central angle.

Proposition V.12.17. *Given is a circle with center O and a chord AB that is not a diameter. A point S on the circle which lies in the interior of the central angle $\angle AOB$ lies on the short arc, and a point L on the circle which lies in the exterior of the central angle lies on the long arc of chord AB.*

Proposition V.12.18. *The perpendicular t to a radius OT, erected at point T, has only one point with a circle in common. The circle lies on one side of this line.*

Definition V.12.6 (Tangent to a circle, an ad hoc definition). The perpendicular erected at a point of the circle to the radius is called the *tangent* of the circle at this point.

Proposition V.12.19. *A line has only one point in common with a circle if and only if it is a tangent.*

Theorem V.12.3. *Given is a circle around O, a point P on the circle, and a line l through point P. The following three statements are equivalent. All three can be used as a definition of the tangent to a circle.*

The line l is perpendicular to the radius OP.

The point P is the only common point of the line and the circle.

The line is the limiting position of a secant PQ in the limit $Q \to P$.

Proposition V.12.20. *The circle-circle intersection property implies the line-circle intersection property.*

Proposition V.12.21. *Assume that the line circle intersection property holds.*

(i) *A line that contains a point inside a circle intersect it in exactly two points.*

(ii) *Let P be a point inside circle $\mathcal{C}$ and Q a point outside the circle. Then the segment PQ intersects the circle in exactly one point, and the line PQ intersects the circle in two points.*

Proposition V.12.22. *Let $\mathcal{C}$ be a circle with center C and radius c, and $\mathcal{D}$ be a circle with center D and radius d.*

(i) *The interiors of the two circles $\mathcal{C}$ and $\mathcal{D}$ are disjoint if and only if $|CD| \geq c + d$.*

(ii) *The interior of circle $\mathcal{C}$ is a subset of the interior of circle $\mathcal{D}$ if and only if $|CD| \leq d - c$.*

(iii) *The interior of circle $\mathcal{D}$ is a subset of the interior of circle $\mathcal{C}$ if and only if $|CD| \leq c - d$.*

(iv) *The interiors of the two circles intersect, but neither one is a subset of the other if and only if $\max\{d - c, c - d\} < |CD| < c + d$.*

In that case, each circle intersects both the interior and exterior of the other one. The interior of each circle intersects the exterior of the other circle.

With inequalities and equalities interpreted as segment comparison and congruence. This Proposition holds in any Hilbert plane, without assuming *neither the Archimedean axiom nor the circle-circle intersection property, nor the circle-line intersection property.*

Definition V.12.7 (Line-circle intersection property). A line that contains a point inside a circle does intersect the circle.

Definition V.12.8 (Circle-circle intersection property). Given are two circles $\mathcal{C}$ and $\mathcal{D}$. Let P, Q be two points on circle $\mathcal{D}$. Assume that point P lies inside circle $\mathcal{C}$ and Q lies outside circle $\mathcal{C}$. Then the two circles $\mathcal{C}$ and $\mathcal{D}$ intersect each other.

Construction V.12.1 (A neutral construction of the tangent to a circle). *Let point A be the intersection of the segment OP with the given circle $\mathcal{C}$. At point A, we erect the perpendicular p onto segment OP. We need a second circle around O through point P. Let B be an intersection point of this second circle with the perpendicular p.*

Finally, we need an intersection point T of circle $\mathcal{C}$ with segment OB. This intersection point is the touching point of a tangent from point P to the given circle $\mathcal{C}$.

Theorem V.12.4 (The in-circle). *All three interior angular bisectors of a triangle intersect in one point. This point has congruent distances from all three sides of the triangle, and hence it is the center of the in-circle.*

Theorem V.12.5 (Farkas Bolyai). *Assume only the axioms of incidence, order, congruence—leading to neutral geometry. If every triangle has a circum-circle, then the Euclidean parallel axiom holds.*

Theorem V.12.6. (a) *Any two equidecomposable figures have the same area.*

(b) *Any two equicomplementable figures have the same area.*

Definition V.12.9 (Pythagorean plane). A *Pythagorean plane* is a Hilbert plane for which the Euclidean parallel postulate holds.

Proposition V.12.23 (In a Pythagorean plane, parallels imply congruent z-angles). [Euclid I.29] *In a Pythagorean plane, any two parallel lines form congruent z-angles with any transversal. Referring to the pair of z-angles shown in the figure on page 244*

$$a \parallel b \Rightarrow \alpha \cong \beta$$

Lemma V.12.2 (Proclus' Lemma). *In any incidence plane where Hilbert's Parallel Postulate for plane geometry holds, the following statements are valid:*

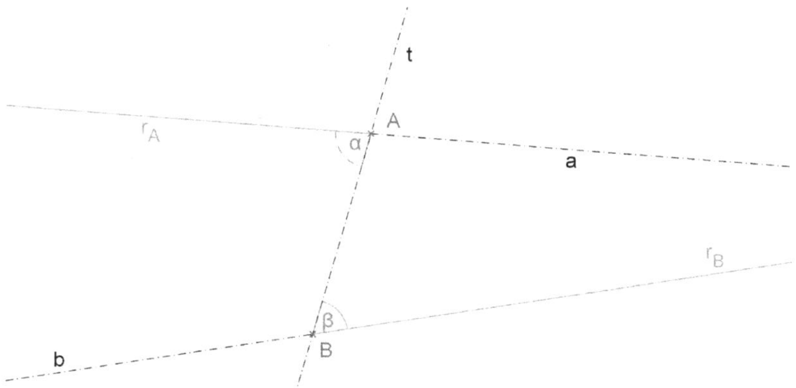

Figure V.12.2: A pair of z-angles.

- *a third line intersecting one of two parallel lines intersects the other one, too.*
- *a third line parallel to one of two parallel lines, is parallel to the other one, too.*

Definition V.12.10 (Traditional Euclidean tools). Constructions with *traditional Euclidean tools* are constructions using only straightedge and compass, done in a finite number of steps of the following types:

1. drawing a line through two different points
2. drawing a circle with a given center through a given point
3. intersecting two lines
4. intersecting a circle and a line
5. intersecting two circles
6. choosing an arbitrary point
7. choosing a point on a given line
8. choosing a point on a given circle
9. deciding whether two points are equal or different

Main Theorem 6 (Uniformity Theorem). *Any Hilbert plane is of either one of the following three types:*

semi-Euclidean *The angle sum of every triangle is two right angle, and every Lambert or Saccheri quadrilateral is a rectangle.*

semi-hyperbolic *The angle sum of every triangle is less than two right angle, and every Lambert or Saccheri quadrilateral has one respectively two acute angles.*

semi-elliptic *The angle sum of every triangle is larger than two right angle, and every Lambert or Saccheri quadrilateral has one respectively two obtuse angles.*

Definition V.12.11 (Aristole's Angle Unboundedness Axiom). For any acute angle θ and any segment PQ, there exists a point X on one side of the angle such that the perpendicular XY dropped onto the other side of the angle is longer than the given segment: $XY > PQ$.

Main Theorem 7 (Proclus' Theorem). *In any Hilbert plane, the Euclidean parallel postulate is equivalent to the following two requirements:*

- *Every triangle has angle sum two right angles.*
- *For every acute angle and every given segment, there exists a right triangle with the given angle and the leg across from it longer than the given segment.*

V.13 Other Axiomatic Systems for Geometry

V.13.1 A Simplified Axiomatization of Geometry, my Suggestion

Introduction

The axiomatic system of Hilbert shows exceptional insight. On the other hand, it is unnatural in many respects. My simplification restricts attention to two dimensions, introduces circles in the natural classical way, and continuity in the way directly related to proofs. My goal is to give a logical analysis of the natural way to do geometry, and still keep up the level of rigor introduced by Hilbert. But I do not try to minimize the axioms as strictly as Hilbert. Especially the axioms dealing with the order of three points on a line, the SAS congruence, and the angle sum can be stated equivalently, but making less assumptions.

0. Undefined elements and relations

Elements:

- A class of undefined objects called *points*, denoted by $A, B, C, \ldots$.
- A class of undefined objects called *lines*, denoted by $a, b, c, \ldots$.
- A class of undefined objects called *circles*, denoted by $\mathcal{C}, \mathcal{D}, \ldots$.

Relations:

- *Incidence* (being *incident, lying on, containing*)
- *Order* (lying *between*) (for points on a line)
- *Congruence* (for segments)
- *Center* (for circle)

I. Axioms of incidence

I.1 For two points A and B there exists a line that contains both points.

I.2 For two different points A and B there exists no more than one line that contains both points.

I.3a Any line contains at least two points.

I.3b There exist at least three points that do not lie on a line.

Remark. I did separate axiom I.3 into I.3a and I.3b, in order to stress there is no direct logical connection between the two sentences intended.

II. Axioms of order

II.1 If a point B lies between a point A and a point C, then the points A, B, C are three distinct points of a line, and B lies between C and A.

Definition V.13.1 (Segment). Let A and B be two distinct points. The *segment* AB is the set consisting of the points A and B and all points lying between A and B. The points A and B are called the *endpoints* of the segment, they are assumed to be different. The points between the endpoints are called the *interior points* of the segment, and the remaining points on the line AB are called the *exterior* points of the segment.

Definition V.13.2 (Ray). Given two distinct points A and B, the *ray* $\overrightarrow{AB}$ is the set consisting of the points A and B, the points inside the segment AB, and all points P on the line AB such that point B lies between vertex A and P. The point A is called the *vertex* of the ray.

II.2* Of any three different points on a line exactly one lies between the two others.

II.3* For two points A and B, the ray $\overrightarrow{AB}$ contains a point C not lying inside the segment AB.

Definition V.13.3 (Triangle). We define a *triangle* to be union of the three segments AB, BC and CA. The three points A, B and C are assumed not to lie on a line. These three points are the *vertices*, and the segments BC, AC, and AB are the *sides* of the triangle.

II.4* (strong Pasch' Axiom) Given is a triangle $\triangle ABC$ and a line a which does not meet any of its vertices A, B, C. If the line a passes through a point of the side AB, it also passes through a point of either one of the two other sides AC or BC, and intersects exactly two sides of the triangle.

II.5 (Convention for orientation) There exist three points A, B and C such that point C lies in the left half-plane of ray $\overrightarrow{AB}$, and the triangle $\triangle ABC$ is oriented counterclockwise.

III. Axioms for circles

III.1 For any center O and point $A \neq O$, there exists a unique circle with center O through the point A.

III.2 A circle $\mathcal{C}$ and a ray emanating from its center O intersect at exactly one point.

Definition V.13.4 (Interior and exterior of a circle). A point A lies *interior* of a circle if it lies between the center O and the point P where the ray $\overrightarrow{OA}$ cuts the circle.

A point B lies *exterior* of a circle if it lies outside the segment OQ between the center O and the point Q where the ray $\overrightarrow{OA}$ cuts the circle.

III.3 (Circle-circle intersection property) Let $\mathcal{C}$ and $\mathcal{D}$ be two circles. Let P, Q be two points on circle $\mathcal{D}$. Assume that point P lies inside circle $\mathcal{C}$ and Q lies outside circle $\mathcal{C}$. Then the two circles $\mathcal{C}$ and $\mathcal{D}$ intersect each other.

III.4 Under the assumptions of (III.3), the two circles intersect in exactly two points, which lie on different sides of the line connecting their centers.

IVa. Axioms of segment congruence

IV.1 If a segment $A'B'$ and a segment $A''B''$ are congruent to the same segment AB, then segment $A'B'$ is also congruent to segment $A''B''$.

IV.1b For any two different points A and B, the segments AB and BA are congruent.

IV.2 (The sums of congruent segments are congruent.) On a line, let AB and BC be two segments which except for B have no point in common. Furthermore, on the same or another line a', let $A'B'$ and $B'C'$ be two segments which except for B' also have no point in common. In that case,

$$\text{if} \quad AB \cong A'B' \quad \text{and} \quad BC \cong B'C' \,, \quad \text{then} \quad AC \cong A'C'$$

IV.3a For a circle $\mathcal{C}$ with center O and two rays with vertex O which cut the circle at points P and Q, the segments OP and OQ are congruent.

IVb. SAS-axiom for triangle congruence

Definition V.13.5 (Angle). An *angle* is the union of two rays with common vertex not lying on one line.

We introduce a segment of *unit length*.

Definition V.13.6 (Congruences of angles). The two angles $\angle POQ$ and $\angle P'O'Q'$ are called *congruent* if and only if there are four units segments $OP, OQ, O'P', O'Q'$ from the vertices and two congruent segments PQ and $P'Q'$ between the sides of the angles.

IV.4* Two triangles ABC and $A'B'C'$ are congruent if they have a pair of congruent angles and the two pairs of adjacent sides are pairwise congruent. In detail:

The congruences $AB \cong A'B'$, $AC \cong A'C'$, $\angle BAC \cong \angle B'A'C'$
imply the congruences $BC \cong B'C'$, $\angle ABC \cong \angle A'B'C'$, $\angle BCA \cong \angle B'A'C'$

V. Axioms of continuity

V.1 (Axiom of Archimedes) If AB and CD are any segments, then there exists a natural number n such that n segments congruent to CD constructed contiguously from A, along a ray from A through B, will pass beyond B.

V.2 (Cantor's principle of boxed intervals) Every sequence of boxed intervals contains a common point. In detail:

For a sequence of segments A_iB_i such that

$$A_i * A_{i+1} * B_{i+1} * B_i \quad \text{for all} \quad i = 1, 2, 3, \ldots$$

there exists a point X^* such that

$$A_i * X^* * B_i \quad \text{for all} \quad i = 1, 2, 3, \ldots$$

VI. Axioms of parallelism

VI.1 (The angle sum of a triangle is two right angles) The sum of the three interior angles of a triangle is two right angles.

VI.2 (Aristotle's Angle Unboundedness Axiom) For any acute angle θ and any segment PQ, there exists a point X on one side of the angle such that the perpendicular XY dropped onto the other side of the angle is longer than the given segment: $XY > PQ$.

Remark. Based on Proclus commentaries to Euclid, Greenberg has suggested the angle unboundedness axiom. This axiom goes back to Aristotle's book I of the treatise *De Caelo* ("On the heavens").

Remark. As shown in volume III, in every Hilbert plane, the Archimedean Axiom implies Aristotle's Axiom.

Problem V.13.1. *Explain, how these axioms imply that each segment is congruent to itself.*

Problem V.13.2. *Explain why congruence of segments is an equivalence relation.*

Problem V.13.3. *Prove that the center of a circle is unique.*

Problem V.13.4. *Prove that every angle is congruent to itself, and to the angle with its sides switched. Explain why congruence of angles is an equivalence relation.*

V.13.2 Birkhoff's Set of Postulates

In 1932, G. D. Birkhoff created a set of four postulates of Euclidean geometry sometimes referred to as Birkhoff's axioms. These postulates are all based on basic geometry that can be confirmed experimentally with a scale and protractor. Since the postulates build upon the real numbers, the approach is similar to a model-based introduction to Euclidean geometry.

Birkhoff's axiom system was utilized in the secondary-school text *Basic Geometry*. Birkhoff's axioms were also modified by the *School Mathematics Study Group* to provide a new standard for teaching high school geometry, known as SMSG axioms.

(I) Postulate of Line Measure. A set of points $\{A, B, \dots\}$ on any line can be put into a one to one correspondence with the real numbers $\{a, b, \dots\}$ such that
$|b - a| = d(A, B)$ for all points A and B.

(II) Point-Line Postulate. There is one and only one line l that contains any two given distinct points P and Q.

(III) Postulate of Angle Measure. A set of rays $\{l, m, n, \dots\}$ through any point O can be put into one to one correspondence with the real numbers $(\mod 2\pi)$ such that for any points $A \neq O$ on line l and $B \neq O$ on line m, the difference $a_m - a_l(\mod 2\pi)$ of the numbers associated with the lines l and m is the angle $\angle AOB$.

Furthermore, if the point B on m varies continuously in a line r not containing the vertex O, the number a_m varies continuously, too.

(IV) Postulate of Similarity. Given are two triangles $\triangle ABC$ and $\triangle A'B'C'$ for which we assume

$$d(A', B') = k \cdot d(A, B)\,,\ d(A', C') = k \cdot d(A, C) \text{ and } \angle B'A'C' = \pm\angle BAC$$

with some constant $k > 0$. Then

$$d(B', C') = k \cdot d(B, C)\,,\ \angle C'B'A' = \pm\angle CBA \text{ and } \angle A'C'B' = \pm\angle ACB$$

V.14 A Synopsis of Euclid

Only the very important items are explained below.

V.14.1 Book I

Definitions

1. A *point* is that which has no part.

2. A *line* is length without breadth.

4. A *straight line* lies evenly with its points.

8. A *plane angle* is the inclination of two lines.

10. When two adjacent angles are equal, they are *right angles.*

15. A *circle* is a line all of whose points are equidistant from a point.

20. A triangle with two equal sides is *isosceles.*

23. *Parallel* straight lines are lines in the same plane that do not meet, no matter how far extended in either direction.

Postulates

1. To draw a line through two points.

2. To extend a given line.

3. To draw a circle with given center through a given point.

4. All right angles are equal.

5. If a line crossing two other lines makes the sum of the interior angles on the same side less than two right angles, then these two lines will meet on that side when extended far enough.

Common Notions

1. Things equal to the same thing are equal to each other.

2. Equals added to equals are equal.

3. Equals subtracted from equals are equal.

4. Things which coincide are equal.

5. The whole is greater than the part.

Propositions

1. To construct an equilateral triangle on a given segment.

2. To draw a segment equal to a given segment at a given point.

3. To cut off a smaller segment from a larger segment.

4. Side-angle-side (SAS) congruence for triangles.

5. The base angles of an isosceles triangle are equal.

6. If the base angles are equal, the triangle is isosceles.

7. It is not possible to put two triangles with equal sides on the same side of a segment.

8. Side-side-side (SSS) congruence for triangles.

9. To bisect an angle.

10. To bisect a segment.

11. To construct a perpendicular to a line at a given point on the line.

12. To drop a perpendicular from a point to a line not containing the point.

13. A line standing on another line makes angles with sum equal to two right angles.

15. Vertical angles are equal.

16. The exterior angle of a triangle is greater than either opposite interior angle.

17. Any two angles of a triangle have a sum less than two right angles.

18. If one side of a triangle is greater than another, then the angle opposite to the greater side is greater than the angle opposite to the smaller side.

19. If one angle of a triangle is greater than another, then the side opposite to the greater angle is greater than the side opposite to the smaller angle.

20. The sum of any two sides of a triangle is greater than the third side.

22. To construct a triangle from all three sides given— provided the sum of any two sides is greater than the remaining side.

23. To produce a given angle at a given point and side.

24. For two triangles with two pairs of equal sides, the triangle with the greater angle between them has the greater opposite side.

25. For two triangles with two pairs of equal sides, the triangle with the greater third side has the greater angle between the two matched sides.

26. Angle-side-angle (ASA) and angle-angle-side (AAS) congruence for triangles.

27. If the two alternate interior angles between a transversal and two lines are equal, then these two lines are parallel.

28. If the exterior angle between a transversal and the first line and the opposite interior angle between the transversal and the second line have sum two right angles, then the two lines are parallel.

28a. If the two interior angles between a transversal and two lines have sum two right angles, then the two lines are parallel.

29. A transversal of two parallel lines makes equal alternate interior angles with the two parallels.

30. Lines parallel to the same line are parallel.

31. To draw a line parallel to a given line through a given point.

32. The sum of angles of a triangle is two right angles.

32a. An exterior angle equals the sum of the two opposite interior angles of a triangle.

33. The lines joining the endpoints of two parallel segments are parallel and the segments cut out by the first two parallels are congruent.

34. The opposite sides and angles of a parallelogram are equal.

35. Parallelograms on the same base and in the same parallels have equal areas.

36. Parallelograms on equal bases and in the same parallels have equal areas.

37. Triangles on the same base and in the same parallels have equal areas.

38. Triangles on equal bases and in the same parallels have equal areas.

39. Triangles with equal areas on the same base lie in the same parallels.

40. Triangles with equal areas on equal bases lie in the same parallels.

41. The area of a parallelogram is twice the area of a triangle on the same base and in the same parallels.

42. To construct a parallelogram with a given angle and area equal to the area of a given triangle.

43. The triangles on opposite sides of a diagonal of a parallelogram are equal.

44. To construct a parallelogram with given side and angle and area equal to the area of a given triangle.

45. To construct a parallelogram with a given angle and area equal to the area of a given figure.

46. To construct a square on a given segment.

47. (Theorem of Pythagoras) The square on the hypothenuse is equal in area to the sum of the squares on the legs of the triangle.

48. If the sum of the areas of squares on two sides of a triangle is equal to the area of the square on the third side, then the triangle is right.

V.14.2 Book II. Propositions

1. The rectangle contained by two lines is the sum of the rectangles contained by one and the segments of the other.

4. The area of the square on the whole segment is equal to the sum of the areas on the two parts of the segment plus twice the area of the rectangle on these two parts.

5. The area of the square on the first longer part of a segment is equal to the area of the rectangle with the difference of the parts and the entire segment as sides, plus the area of the square on the second part. [15]

6. The area of a rectangle with one side a longer segment and an added piece, the second side the added piece plus the area of the square on half of the longer segment add up to the area of the square on the half plus the added piece. [16]

11. To cut a longer segment so that the rectangle on the whole and the part is equal to the square on the other part. [17]

14. To construct a square with area equal to that of a given figure.

V.14.3 Book III. Propositions

1. To find the center of a circle.

2. The segment joining two points of a circle lies inside the circle.

5. If two circles intersect, they do not have the same center.

6. If two circles are tangent, they do not have the same center.

10. Two circles can intersect in at most two points.

11. 12. If two circles are tangent, their centers lie in a line with the point of tangency.

16. The line perpendicular to a diameter at its endpoint is tangent to a circle, and the "angle between the tangent line and the circle" [18] is less than any "rectilineal angle".

---

[15] $p^2 = [p - (c-p)]c + (c-p)^2$

[16] $(c+p)p + \left(\frac{c}{2}\right)^2 = \left(\frac{c}{2} + p\right)^2$

[17] Solve $cx = (c-x)^2$ for the part x.

[18] The notion in quotes is not well defined by today's standard.

17. To draw a tangent to a circle from a point outside the circle.

18. A tangent line to a circle is perpendicular to the radius at the point of tangency.

19. The perpendicular to a tangent line at the point of tangency will pass through the center of a circle.

20. The angle at the center is twice the angle at a point of the circumference subtending a given arc of a circle.

21. Two angles from points of a circle subtending the same arc are equal.

22. The opposite angles of a quadrilateral in a circle have sum equal to two right angles.

31. The angle in a semicircle is a right angle.

32. The angle between a tangent line and a chord of a circle is equal to the angle on the arc cut off by the chord.

35. If two chords cut each other inside the circle, the rectangle on the segments on one chord is equal in area to the rectangle on the segments on the other chord.

36. From a point outside a circle, let a tangent and a secant be drawn. The square on the segment on the tangent is equal in area to the rectangle formed by the two segments from the point outside of the circle to the endpoints of the secant.

37. From a point outside a circle are drawn two lines cutting or touching the circle. If the square on the segment on one line is equal in area to the rectangle formed by the two segments from the point outside to the endpoints of the secant on the other line, then the first line is tangent to the circle.

V.14.4 Book IV. Propositions

1. To inscribe a given segment in a circle.

2. To inscribe a triangle, equiangular to a given triangle, in a circle.

3. To circumscribe a triangle, equiangular to a given triangle, around a circle.

4. To inscribe a circle in a triangle.

5. To circumscribe a circle around a triangle.

10. To construct an isosceles triangle whose base angles are twice the vertex angle.

11. To inscribe a regular pentagon in a circle.

12. To circumscribe a regular pentagon around a circle.

15. To inscribe a regular hexagon in a circle.

16. To inscribe a regular 15-sided polygon in a circle.

V.14.5 Book V

Definitions

4. Magnitude are said to *have a ratio* if either one, being multiplied, an exceed the other.

5. Four magnitudes a, b, c, d are in the same ratio $a : b$ equal to $c : d$, if and only if for any whole numbers m and n either one of these three cases occurs:

$$\begin{array}{lll} \text{both} & ma > nb \text{ and } mc > nd, \\ \text{both} & ma = nb \text{ and } mc = nd, \\ \text{both} & ma < nb \text{ and } mc < nd. \end{array}$$

V.14.6 Book VI. Propositions

1. The areas of triangles of the same height are in the same ratio as their bases.

2. A line is parallel to the base of a triangle if and only if it cuts the two other sides in the same ratio—proportionately.

3. A line from the vertex of a triangle to the opposite side bisects the angle if and only if it cuts the opposite side in proportion to the remaining sides of the triangle.

4. The sides of equiangular triangles are proportional.

5. If the sides of two triangles are proportional, their corresponding angles are pairwise equal.

6. If two triangles have one pair of equal angles, and the sides containing these pair are proportional, then the triangles are similar.

8. The altitude from the right angle of a right triangle divides it into two triangles similar to each other and to the whole triangle.

12. To find a fourth proportional for three given segments.

13. To find a mean proportional between two given segments.

16. Four segments are proportional if and only if the rectangle on the extremes is equal in area to the rectangle on the means.

30. To cut a segment in extreme and mean ratio. [19]

31. Any figure on the hypothenuse of a right triangle has area equal to the sum of the areas of similar figures on the legs of the triangle.

V.14.7 Book X

1. Given two unequal quantities, if one subtracts from the greater one the smaller one, and repeats this process enough times, there will remain a quantity lesser than the smaller of the two original quantities.

117. [20] The diagonal of a square is incommensurable with its side.

V.14.8 Book XI

Definitions

25. A *cube* is a polyhedron made of six equal squares.

26. An *octahedron* is a polyhedron made of eight equal equilateral triangles.

27. An *icosahedron* is a polyhedron made of twenty equal equilateral triangles.

28. An *dodecahedron* is a polyhedron made of twelve equal regular pentagons.

---

[19] Solve $c : x = x : p$ for x.

[20] not in Heath, but in Commandino

Propositions

21. The plane angles in a solid angle make a sum less than four right angles.

28. A parallelepiped is bisected by its diagonal plane.

29. 30. Parallelepipeds on the same base and of the same height are equal in volumes.

31. Parallelepipeds on equal bases and of the same height are equal in volumes.

V.14.9 Book XII. Propositions

2. The areas of circles are in the same ratio as the squares on their diameters.

3. A pyramid is divided into two pyramids and two prisms.

5. Pyramids of the same height on triangular bases have volumes in the ratio of the areas of the bases.

6. A prism with triangular bases is divided into three triangular pyramids with equal volumes.

V.14.10 Book XIII. Propositions

7. If at least three angles of an equilateral pentagon are equal, the pentagon is regular.

10. In a circle, the square on the side of the inscribed pentagon is equal in area to the square on the side of the inscribed hexagon plus the square on the inscribed decagon. [21]

13. To inscribe a tetrahedron in a sphere.

14. To inscribe a octahedron in a sphere.

15. To inscribe a cube in a sphere.

16. To inscribe an icosahedron in a sphere.

17. To inscribe a dodecahedron in a sphere.

18. (Postscript) Besides these five figures there is no other contained by equal regular polygons.

---

[21] $a_5^2 = R^2 + a_{10}^2$

Problem V.14.1. *Which ones of these sentences are easily misinterpreted, unclear or even wrong?*

V.15 The History of Translations and Editions of Euclid's Elements

"Jüngling, merke dir in Zeiten,

Wo sich Geist und Sinn erhöht:

Dass die Muse zu g e l e i t e n ,

Doch zu l e i t e n nicht versteht."

—*Johann Wolfgang von Goethe*

Young man, remember in times,
When mind and spirits rise:
That to accompany the muse can,
But conducting does not understand.

—*Johann Wolfgang von Goethe*

The highly acclaimed *Elements* is simply a huge collection—divided into 13 books—of 465 propositions from plane and solid geometry. Today, it is generally agreed that relatively few of these theorems were Euclid's own invention. Rather, by compiling, editing, and perfecting the known body of Greek mathematics, he created a superbly organized treatise. The *Elements* were so successful and revered that they thoroughly obliterated all preceding works of its kind. They have become a lasting contribution to mankind.

It is a misfortune that no copy of Euclid's *Elements* has been found that actually dates from the author's own time. Hence his writings have had to be reconstructed from numerous recensions, commentaries, and remarks by other writers. Proclus makes it clear that the *Elements* were highly valued in Greece and refers to numerous commentaries on it. Among the most important must have been those by Heron (100 B.C. - A.D.100), Porphyry and Pappus (third century A.D.).

Modern editions of the work are based upon a revision that was prepared by the Greek commentator Theon of Alexandria (end of the fourth century A.D). He lived almost 700 years after the time of Euclid. Until 1808, Theon's revision was the oldest edition of the *Elements* known. In 1808, Napoleon ordered valuable manuscripts to be taken from Italian libraries and sent to Paris. At that occasion, Francois Peyrard found, in the Vatican library, a tenth-century copy of an edition of Euclid's *Elements* that predates Theon's recension. The historians J.L.Heiberg and Thomas L. Heath have used principally this manuscript. A study of this older edition and a careful sifting of citations and remarks made by early commentators indicate that the introductory material of Euclid's original treatise undoubtedly underwent some editing in the subsequent revisions. But the propositions and their proofs have remained as Euclid wrote them, except for minor additions and deletions.

There are also Arabic translations of Greek works and Arabic commentaries, presumably based on Greek manuscripts no longer available. The first really satisfying Arabic translation of the *Elements* was done by Tâbit ibn Qorra (826-901). These, too, have been used to decide what was in Euclid's original. But the Arabic translations and revisions are on the whole inferior to the Greek manuscripts.

The first complete Latin translation of the *Elements* were not made from the Greek but from Arabic. In the eighth century, a number of Byzantine manuscripts of Greek works were translated by the Arabians. In 1120, the English scholar Adelard of Barth (ca. 1075-1160) made a Latin translation of the *Elements* from one of the older Arabian translations. Other Latin translations were made from the Arabic by Gherardo of Cremona (1114-1187) and, 150 years after Adelard, by Johannes Campanus.

Euclid's *Elements* were widely read, translated and changed by people all over the world. From 1601 to 1607, Mattes Ricci (1552-1610) from Italy and Hsü Kuang-ching (1562-1634) translated the first six books into Chinese. This played a significant role in the subsequent development of mathematics in China.

The first printed edition of the *Elements* was made at Venice in 1482 and contained Campanus' translation. This very rare book was beautifully executed. It was the first mathematical book of any consequence to be printed. In 1574, Clavius (1537-1612) published an edition that is valuable for its extensive scholia. An important Latin translation from the Greek was made by Commandino in 1572. This translation served as a basis for many subsequent translations, including the very influential work by Robert Simson. From the latter, many English editions were derived. The first complete English translation of the *Elements* was the monumental Billingsley translation issued in 1570.

Nicolaus Mercator (ca. 1620-1687) published an edition, along with his works in trigonometry, astronomy and other topics. Adrien-Marie Legendre (1752-1833) is known for his very popular *Éléments de géométrie*, in which he attempted a pedagogical improvement of Euclid's *Elements* by considerably rearranging and simplifying many of the propositions. He used some algebra not in the original. The high school versions most widely used during our century are patterned on Legendre's modification of Euclid's work.

Euclid's *Elements* is to geometry what the Bible is to Christianity. No work, except the Bible, has been more widely used, edited, or studied than Euclid's *Elements*. Over one thousand editions of Euclid's *Elements* have appeared since the first one printed in 1482. For more than two millennia, this work has dominated all teaching of geometry. Abraham Lincoln is credited with learning logic by studying Euclid's *Elements*, as his biographer Carl Sandburg tells. In his autobiography, Bertrand Russell (1872-1970) penned this remarkable recollection: "At the age of eleven, I began Euclid, with my brother as tutor. This was one of the great events of my life, as dazzling as first love." Probably no work has exercised a greater influence on scientific thinking.

Bibliography

[1] Solomon Feferman Anita B. Feferman, *Alfred Tarski: Life and Logic*, Cambridge University Press, 2004.

[2] H.S.M. Coxeter and S.L. Greitzer, *Geometry Revisited*, ninth printing, New Mathematical Library 19, The Mathematical Asociation of America, 1967.

[3] H.E. Dudeney, *The Canterbury Puzzles, and other curios problems*, Nelson, London, 1929.

[4] F.R.Moulton, *A simple non-desarguesian plane geometry*, Trans. Math. Soc. (1902).

[5] Marvin J. Greenberg, *Euclidean and Non-Euclidean Geometry*, fourth ed., W.H. Freeman and Company, New York, 2008.

[6] ______, *Old and new results in the foundations of elementary plane Euclidean and Non-Euclidean geometries*, The American Mathematical Monthly **117** (2010), 198–219.

[7] Robin Hartshorne, *Geometry: Euclid and Beyond*, second printing, Springer, 2002.

[8] ______, *Non-Euclidean iii.36*, The American Mathematical Monthly **110** (2003), 495–502.

[9] David Hilbert, *Foundations of Geometry*, 2nd English ed., Open Court, La Salle, 1971.

[10] ______, *Grundlagen der Geometrie*, 14. Auflage, B.G.Teubner, Stuttgart, Leibzig, 1999.

[11] Ross Honsberger, *Episodes in nineteenth and twentieth century Euclidean geometry*, New Mathematical Library, vol. 37, The Mathematical Association of America, 2005.

[12] Roger A. Johnson, *Advanced Euclidean Geometry*, second printing, Dover Publications, Inc., Mineola, New York, 2007.

[13] Wilbur Richard Knorr, *The Ancient Tradition of Geometric Problems*, Dover Publications, 1945/1993.

[14] Stefan Mykytiuk and Abe Shenitzer, *Four significant axiomatic systems and some of the issues associated with them*, The American Mathematical Monthly **102** (1995), 62–67.

[15] David Park, *The Grand Contraption— the world in myth, number, and chance*, second printing, Princeton University Press, 2005.

[16] Alexander Ryba and Joseph Stern, *Equimodular polynomials and the tritangency theorems of Euler, Feuerbach and Guinand*, The American Mathematical Monthly **118** (2011), 217–228.

[17] James R. Smart, *Modern Geometries*, fifth ed., Brooks/Cole Publishing Company, 1997.

[18] John Stillwell, *The Four Pillars of Geometry*, Springer, 2005.

[19] B.L. van der Waerden, *Science Awakening I*, Kluwer Academic Publishers, 1954.

[20] Robert S. Wolf, *A Tour through Mathematical Logic*, the Carus mathematical monographs 30, Mathematical Association of America, 2005.

Index

A Course in Old and New Geometry

Volume II: Basic Euclidean Geometry

The *front cover* shows the figure from page 113 about the geometric proof of the law $a(bc) = b(ac)$. This line of reasoning does not presuppose Pappus' theorem, instead we use properties of circles in a Pythagorean plane.

A decade long experience of teaching the course *Fundamental of Geometry* at the University of North Carolina at Charlotte, many notes for exercises, reading and research are the bases for this bulky work. Retired I could find the time and leisure to prepare several volumes. My heartfelt thanks go to everyone involved, this means famous authors as well as my former students.

The present second volume starts with Euclidean geometry, and the text gets a more educational and even more elementary flavour. Here I start with Thales theorem about the angle in a semicircle, and continue with Euclid's related theorems about angles in a circle. The most simple parts of the Euclidean geometry are given in detail, as well as the later parts about similarity, and finally area in Euclidean geometry with the theorem of Pythagoras and trigonometry, and the measurement of the circle. Too, the lens equation from geometrical optics is treated.

In several sections is recalled the strictly modern view of the first volume, beginning with Hilbert's axioms from the *Foundations of Geometry* [9]. After some discussion of logic and axioms in general, we go on with a short section about incidence geometries in two dimensions. The theorems from neutral geometry which have been proved in the first volume and are needed again, are cited. The relation of analytic and synthetic geometry is treated on Hilbert's rigorous account.

Franz Rothe has received his doctoral degree in mathematics from the university of Tübingen, Germany. He has been professor at the University of North Carolina at Charlotte, and has published about 40 articles and a lecture notes in mathematics, and further books on geometry, number theory, modern algebra, and graph theory. Dr. Rothe is retired since several years, and is now emeritus professor.

www.ingramcontent.com/pod-product-compliance
Lightning Source LLC
LaVergne TN
LVHW080037210926
848481LV00011B/3214
9798231094943